MASSEY-FERGUSON

SHOP MANUAL

Information and Instructions

This shop manual contains several sections each covering a specific group of wheel type tractors. The Tab Index on the preceding page can be used to locate the section pertaining to each group of tractors. Each section contains the necessary specifications and the brief but terse procedural data needed by a mechanic when repairing a tractor on which he has had no previous actual experience.

Within each section, the material is arranged in a systematic order beginning with an index which is followed immediately by a Table of Condensed Service Specifications. These specifications include dimensions, fits, clearances and timing instructions. Next in order of arrangement is the procedures paragraphs.

In the procedures paragraphs, the order of presentation starts with the front axle system and steering and proceeding toward the rear axle. The last paragraphs are devoted to the power take-off and power lift systems. Interspersed where needed are additional tabular specifications pertaining to wear limits, torquing, etc.

HOW TO USE THE INDEX

Suppose you want to know the procedure for R&R (remove and reinstall) of the engine camshaft. Your first step is to look in the index under the main heading of ENGINE until you find the entry "Camshaft." Now read to the right where under the column covering the tractor you are repairing, you will find a number which indicates the beginning paragraph pertaining to the camshaft. To locate this wanted paragraph in the manual, turn the pages until the running index appearing on the top outside corner of each page contains the number you are seeking. In this paragraph you will find the information concerning the removal of the camshaft.

More information available at Clymer.com
Phone: 805-498-6703

Haynes Publishing Group
Sparkford Nr Yeovil
Somerset BA22 7JJ England

Haynes North America, Inc
859 Lawrence Drive
Newbury Park
California 91320 USA

ISBN 10: 0-87288-539-9
ISBN-13: 978-0-87288-539-4

Common spark plug conditions

NORMAL

Symptoms: Brown to grayish-tan color and slight electrode wear. Correct heat range for engine and operating conditions.
Recommendation: When new spark plugs are installed, replace with plugs of the same heat range.

WORN

Symptoms: Rounded electrodes with a small amount of deposits on the firing end. Normal color. Causes hard starting in damp or cold weather and poor fuel economy.
Recommendation: Plugs have been left in the engine too long. Replace with new plugs of the same heat range. Follow the recommended maintenance schedule.

TOO HOT

Symptoms: Blistered, white insulator, eroded electrode and absence of deposits. Results in shortened plug life.
Recommendation: Check for the correct plug heat range, over-advanced ignition timing, lean fuel mixture, intake manifold vacuum leaks, sticking valves and insufficient engine cooling.

CARBON DEPOSITS

Symptoms: Dry sooty deposits indicate a rich mixture or weak ignition. Causes misfiring, hard starting and hesitation.
Recommendation: Make sure the plug has the correct heat range. Check for a clogged air filter or problem in the fuel system or engine management system. Also check for ignition system problems.

PREIGNITION

Symptoms: Melted electrodes. Insulators are white, but may be dirty due to misfiring or flying debris in the combustion chamber. Can lead to engine damage.
Recommendation: Check for the correct plug heat range, over-advanced ignition timing, lean fuel mixture, insufficient engine cooling and lack of lubrication.

ASH DEPOSITS

Symptoms: Light brown deposits encrusted on the side or center electrodes or both. Derived from oil and/or fuel additives. Excessive amounts may mask the spark, causing misfiring and hesitation during acceleration.
Recommendation: If excessive deposits accumulate over a short time or low mileage, install new valve guide seals to prevent seepage of oil into the combustion chambers. Also try changing gasoline brands.

HIGH SPEED GLAZING

Symptoms: Insulator has yellowish, glazed appearance. Indicates that combustion chamber temperatures have risen suddenly during hard acceleration. Normal deposits melt to form a conductive coating. Causes misfiring at high speeds.
Recommendation: Install new plugs. Consider using a colder plug if driving habits warrant.

OIL DEPOSITS

Symptoms: Oily coating caused by poor oil control. Oil is leaking past worn valve guides or piston rings into the combustion chamber. Causes hard starting, misfiring and hesitation.
Recommendation: Correct the mechanical condition with necessary repairs and install new plugs.

DETONATION

Symptoms: Insulators may be cracked or chipped. Improper gap setting techniques can also result in a fractured insulator tip. Can lead to piston damage.
Recommendation: Make sure the fuel anti-knock values meet engine requirements. Use care when setting the gaps on new plugs. Avoid lugging the engine.

GAP BRIDGING

Symptoms: Combustion deposits lodge between the electrodes. Heavy deposits accumulate and bridge the electrode gap. The plug ceases to fire, resulting in a dead cylinder.
Recommendation: Locate the faulty plug and remove the deposits from between the electrodes.

MECHANICAL DAMAGE

Symptoms: May be caused by a foreign object in the combustion chamber or the piston striking an incorrect reach (too long) plug. Causes a dead cylinder and could result in piston damage.
Recommendation: Repair the mechanical damage. Remove the foreign object from the engine and/or install the correct reach plug.

SHOP MANUAL
MASSEY-FERGUSON

MODELS

340, 350, 355, 360 AND 399

The tractor serial number is located in two locations, on the serial number plate (Fig. 1) located on the right side of the tractor and stamped on the rear axle casting (Fig. 2). The tractor serial number is coded to identify the machine type, year of manufacture, week of manufacture (during specific year) and specific unit during week.

The illustration shown in Fig. 3 identifies a two-wheel-drive 360 tractor which was the 132nd tractor manufactured during the 33rd week of 1988.

Fig. 1—The tractor serial number is located on plate attached to tractor right side as shown as well as stamped on axle housing as shown in Fig. 2.

Fig. 2—Tractor serial number is stamped into rear axle housing as shown as well as on plate shown in Fig. 1. Numbers match on original assembly.

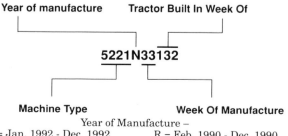

Fig. 3—The tractor serial number identifies machine type, year of manufacture, week of manufacture during specific year, and specific unit manufactured during that week. The illustration shown identifies a two-wheel-drive 360 tractor which was the 132nd tractor manufactured during the 33rd week of 1988.

Year of Manufacture –

A = Jan. 1992 - Dec. 1992	R = Feb. 1990 - Dec. 1990
B = Jan. 1993 - Dec. 1993	S = Jan. 1991 - Dec. 1991
N = Feb. 1988 - Jan. 1989	U = Aug. 1986 - Jan. 1987
P = Feb. 1989 - Jan. 1990	V = Feb. 1987 - Jan. 1988

Machine Type –

5000 M-F 350 2WD	5222 M-F 360 4WD
5001 M-F 350 4WD	5309 M-F 355 2WD
5012 M-F 399 2WD	5310 M-F 355 4WD
5013 M-F 399 4WD	5747 M-F 340 4WD
5221 M-F 360 2WD	5812 M-F 340 4WD

INDEX (By Starting Paragraph)

**Models
340, 350, 355,
360, and 399**

**Model
340, 350, 35
360, and 3**

DUAL DIMENSIONS

This service manual provides specifications in both U.S. Customary and Metric (SI) systems of measurement. The first specification is given in the measuring system perceived by us to be the preferred system when servicing a particular component, while the second specification (given in parenthesis) is the converted measurement. For instance, a specification of 0.011 inch (0.28 mm) would indicate that we feel the preferred measurement in this instance is the U.S. Customary system of measurement and the Metric equivalent of 0.011 inch is 0.28 mm.

CONDENSED SERVICE DATA

	340-350	355	Models 360	Early 399*	Late 399†
GENERAL					
Engine Make			Perkins		
Model	AD3.152S	AT3.1524		A6.3544	1006.6
Build Code	CE31160	CN31187	CN31185	TW31159	YA31240
Number of Cylinders		3		6	
Bore		91.4 mm (3.6 in.)		98.4 mm (3.875 in.)	100 mm (3.937 in.)
Stroke		127 mm (5.0 in.)			
Displacement		2.5 L (152 cid)		5.8 L (354 cid)	6.0 L (365 cid)
Compression Ratio	16:1	15.5:1		16:1	16.5:1
Firing Order		1-2-3		1-5-3-6-2-4	1-5-3-6-2-4
Valve Clearance (Cold) –					
Inlet	0.3 mm (0.012 in.)		0.2 mm (0.008 in.)		
Exhaust		0.3 mm (0.012 in.)			0.45 mm (0.018 in.)
Valve Face Angle –					
Inlet and Exhaust		35°		45°	
Valve Seat Angle –					
Inlet and Exhaust		36°		46°	
Injection timing - BTDC					
Static	16°	16°‡	17°	23°	22°
Fuel Pump –					
Make		CAV			
Model		DPA			
Engine Low Idle, rpm	725-775	725-775‡		725-775	
Engine High Idle, rpm	2470	2370‡	2470	2420#	2310
Engine Rated Speed, rpm		2250		2200	
Battery Terminal Grounded		Negative			
SIZES					
Crankshaft Main Journal Diameter		See Paragraph 58		See Paragraph 87	See Paragraph 118

CONDENSED SERVICE DATA (CONT.)

			Models		
	340-350	**355**	**360**	**Early 399***	**Late 399†**
SIZES (Cont.)					
Crankshaft Crankpin Diameter	—————— See Paragraph 58 ——————			See Paragraph 87	See Paragraph 118
Camshaft Journal Diameter .	—————— See Paragraph 53 ——————			See Paragraph 82	See Paragraph 113
Piston Pin Diameter	—————— See Paragraph 56 ——————			See Paragraph 85	See Paragraph 116
Valve Stem Diameter	—————— See Paragraph 39 ——————			See Paragraph 70	See Paragraph 99
CLEARANCES					
Main Bearing Diametral Clearance	—————— See Paragraph 58 ——————			See Paragraph 87	See Paragraph 118
Rod Bearing Diametral Clearance	—————— See Paragraph 57 ——————			See Paragraph 86	See Paragraph 117
Camshaft Bearing Diametral Clearance	—————— See Paragraph 53 ——————			See Paragraph 82	See Paragraph 113
Crankshaft End Play	—————— See Paragraph 58 ——————			See Paragraph 87	See Paragraph 118
Piston Skirt to Cylinder Clearance	—————— See Paragraph 55 ——————			See Paragraph 84	See Paragraph 115
CAPACITIES					
Cooling System	———— 9.8 L ———— (10.4 qts.)			20.1 L (21.2 qts.)	23 L (24 qts.)
Crankcase With Filter	———— 6.8 L ———— (7.2 qts.)			15.4 L (16.2 qts.)	14.3 L (15.2 qts.)
Transmission - Without Spacer	———————————— 43.4 L ———————————— (11.5 gal.)				
Transmission - With Spacer or Transfer Housing	———————————— 47.4 L ———————————— (12.5 gal.)				
Final Drive Rear Planetary Hub - Each side .	. . . .	. . . .	. . . .	0.9 L (5 pints)	0.9 L (5 pints)
Hydrostatic Steering	———— 0.7 L ———— (1.2 pints)			———— 1.2 L ———— (2.2 pints)	

CONDENSED SERVICE DATA (CONT.)

	340-350	355	Models 360	Early 399*	Late 399†

CAPACITIES (Cont.)

Front Drive Axle Hubs
(Each Side) . . . . ————————1.2 L———————— ————1.3 L————
(2.3 pints) (2.6 pints)

Front Drive
Axle Housing . . . . ————————4.0 L———————— ————5.8 L————
(1.1 gal.) (1.5 gal.)

* M-F 399 tractors before 1990 (serial number S00000).
† M-F 399 tractors beginning with serial number S01001 (1991 and later).
‡ Low idle speed should be 750 rpm, high idle rpm should be 2470 rpm and static timing should be 17° BTDC for M-F 355 models after engine serial number U796441P.
# High idle no load speed should be 2310 rpm for M-F 399 after S. N. U793030P.

FRONT AXLE SYSTEM
(TWO-WHEEL DRIVE)

FRONT AXLE ASSEMBLY AND STEERING LINKAGE

Two-Wheel-Drive Models

1. WHEELS AND BEARINGS. To remove front wheel hub and bearings, raise and support the front axle extension, then unbolt and remove the tire and wheel assembly. Remove cap (2 or 3—Fig. 4), cotter pin (4), castellated nut (5), washer and outer bearing cone (7). Slide the hub assembly from spindle axle shaft. Remove dust shield (12), seal (9) and inner bearing cone (11). Drive bearing cups (8 and 10) from hub if renewal is required. Pack wheel bearings liberally with a multi-purpose lithium based grease. Reassemble by reversing disassembly procedure. Tighten castellated nut (5) to a torque of 80 N·m (60 ft.-lbs.), then back nut off to the nearest hole and install cotter pin (4). Be sure to install cap (2 or 3) securely.

2. TRACK ROD AND TOE-IN. All models are equipped with hydrostatic steering. On 340-360 models, a single track rod connects the left and right steering arms which are attached to the steering spindles. On 399 models, the two track rods are

attached to each end of the hydrostatic steering rod, which is located between the steering arms. The track rod of all models assures that both left and right wheels turn in unison and the distance between ends of track rod establishes front wheel toe-in.

Ends of track rod are automotive type and should be renewed if wear is excessive. The procedure for removing and installing ends is self-evident. Recommended toe-in is 0-5 mm (0-3/16 inch) for all models. Toe-in should be measured between the wheel rims on center line of axle, parallel to ground. Rotate wheels and remeasure to be sure that wheels are not bent giving incorrect reading.

Distance between front wheels is adjustable to seven different widths by relocating axle extensions and changing length of track rod using the pre-existing attachment holes.

On 340-360 models, axle extension and hydrostatic steering cylinder bracket retaining screws (1—Fig. 5) should be tightened to 180-230 N·m (135-170 ft.-lbs.) torque. Tighten track setting screw (2) to 45-55 N·m (33-40 ft.-lbs.) torque before tightening locknut. Note that spacers are used on the outer screws between steering cylinder brackets and axle extensions when axle width is at the four widest settings. To make small toe-in adjustments, remove track adjusting screw (2) from left end of track rod, loosen rod end clamp bolt on right end, then turn center section of track rod until toe-in is correct. Reinstall track adjusting screw (2) and tighten rod end clamp bolt to 45-55 N·m (33-41 ft.-lbs.) torque.

Fig. 4—Exploded view of wheel hub typical of two-wheel-drive models.

2. Hub cap
3. Hub cap
4. Cotter pin
5. Castellated nut and tang washer
6. Hub
7. Outer bearing cone

8. Outer bearing cup
9. Seal
10. Inner bearing cup
11. Inner bearing cone
12. Dust shield
17. Wheel retaining screws

Fig. 5—View of axle left end typical of 340-360 models. Track adjusting screw and lock nut (2) and cylinder bracket and axle extension screws (1) must be properly installed and tightened.

Fig. 6—View of axle left end typical of type used on 399 models. Right side is equipped with a similar track rod, adjusting bolt (2) and rod end. Rod end may be locked with clamp bolt (3) as shown or jam nut.

On 399 models, axle extension retaining screws (1—Fig. 6) should be tightened to 340-450 N•m (250-330 ft.-lbs.) torque. Tighten track setting bolt (2) to 120-160 N•m (90-120 ft.-lbs.). To make small toe-in adjustments, remove track adjusting bolt (2), loosen rod end clamp bolt (3) or jam nut, then turn track rod until toe-in is correct. Reinstall track adjusting bolt (2) and tighten to 120-160 N•m (90-120 ft.-lbs.) torque. Tighten rod end clamp bolt (3) to 45 N•m (33 ft.-lbs.) torque or jam nut to 160-200 N•m (120-130

ft.-lbs.) torque. Equal toe-in adjustments should be made to both sides to center steering.

3. SPINDLES, AXLE EXTENSIONS AND BUSHINGS. To remove spindle (12—Fig. 7 or Fig. 8), first remove the wheel and hub. Disconnect rod end (2) from steering arm (1), remove clamp screw (13) from steering arm, then remove steering arm. Remove key (14) and seal (20) from top of spindle, then lower spindle out of axle extension (19). Remove thrust bearing (11) from spindle. Clean and inspect parts for wear or other damage and renew as necessary.

Each axle extension (19) is equipped with two spindle bushings that must be reamed after pressing into position. Clean all metal particles from bore and be sure that hole for grease fitting is clean and open before assembling.

When reassembling, install thrust bearing (11) on spindle so that numbered side of bearing is facing upward and insert spindle through axle extension. Install seal (20) and key (14) then locate steering arm on top of spindle. Tighten steering arm retaining clamping screw to a torque of 125-165 N•m (94-122 ft.-lbs.) for 340-360 models; 280-370 N•m (207-273 ft.-lbs.) torque for 399 models. Refer to paragraph 2 for track and toe-in adjustment and other recommended torques. Balance of reassembly is the reverse of disassembly.

4. AXLE CENTER MEMBER, PIVOT PIN AND BUSHINGS. To remove front axle assembly, first remove any front mounted equipment, guards, weights and weight frame. Raise front of tractor in such a way that it will not interfere with the removal

Fig. 7—Exploded view of front axle typical of two-wheel-drive 340-360 models.

1. Steering arm
2. Rod end
3. Cylinder pivot pin
4. Axle pivot pin
5. Steering cylinder
6. Track rod
7. Thrust washer
8. Steering cylinder bracket
9. Tapered retaining pin
10. Axle center member
11. Thrust washers
12. Spindle
18. Bushing
19. Axle extension
20. Seal
21. Shims

Fig. 8—Exploded view of front axle typical of type used on two-wheel-drive 399 models.

1. Steering arm
2. Rod end
3. Cylinder ball joint
4. Axle pivot pin
5. Steering cylinder
6. Track rod
7. Thrust washer
8. Pivot block
9. Retaining pin
10. Axle center member
11. Thrust washers
12. Spindle
14. Key
18. Bushing
19. Axle extension
20. Seal
21. Shims
22. Bushing
23. Shims
24. Pivot pin
25. Hydraulic hoses
26. Jam nut

Fig. 9—Axle pivot bushings should be installed as shown for 340-360 models. Hole (B) should be aligned with grease passage and slot (C) should be in position indicated.

of the axle, such as with a support located under engine sump. Removal of wheels, spindles and axle extensions will reduce weight and may make handling the center member easier; however, the complete axle assembly can be removed as a unit. Disconnect hydrostatic steering hoses from the steering cylinder or cylinders and cover openings to prevent the entry of dirt. Support the axle with a suitable jack to prevent tipping while permitting the axle to be lowered and moved safely. Remove retaining pin (9—Fig. 7 or Fig. 8), then use a suitable puller to withdraw axle pivot pin (4). Carefully lower the axle assembly and roll axle from under tractor.

Check axle pivot bushings and renew if necessary. Bushings are located in support housing of 340-360 models and should be installed flush to 0.5 mm (0.020 inch) below flush (A—Fig. 9) with housing bore. Split (Fig. 9) in bushing should be down and hole (B) for grease passage should be up as shown. Axle pivot bushings (22—Fig. 8) are located in axle of 399 models. On all models, it may be necessary to ream bushings after installation. Reverse removal procedure when assembling. Axle end play should be 0.05-0.25 mm (0.002-0.010 inch) on pivot pin. Push the axle toward rear on pivot pin, then measure axle end play with a feeler gauge as shown in Fig. 10 or Fig. 11. Shims (21—Fig. 7 or Fig. 8) are available in various thick-

Fig. 10—End play of front axle should be measured with a feeler gauge as shown for 340-360 model with two-wheel drive.

Fig. 12—Measure gap (A) between the top spacer tube on right side and front support of 340-360 models.

Fig. 11—Measure end play of front axle with a feeler gauge as shown for model 399 with two-wheel drive.

nesses for adjusting end play. On 340-360 models, make sure that tapered pin (9—Fig. 7) is correctly and firmly seated, then tighten retaining nut to 80-140 N·m (70-110 ft.-lbs.) torque. On all models, refer to paragraphs 2 and 3 for additional torque values and assembly notes.

5. FRONT SUPPORT. To remove the front support, the axle must be removed, the radiator must be removed and the front support must be unbolted from the front of engine. The front axle, the front support and the remainder of the tractor must each be supported separately while removing, while separated

and while assembling. Be sure that sufficient equipment is available before beginning.

To remove front axle support from 340-360 models, first remove any front mounted equipment, guards, weights and weight frame. Remove grille, hood, hood side panels, air cleaner and battery. Drain cooling system, disconnect radiator hoses and disconnect wires to headlights. Remove radiator and oil cooler assembly, then refer to paragraph 4 and remove the axle assembly. Attach a hoist or other supporting device to the front support, then unbolt and separate the front support from the front of the engine. Be careful not to lose shims which may be located between front support casting and front of engine.

To remove front axle support from 399 models, first remove any front mounted equipment, guards, weights and weight frame. Remove fuel tank and radiator and radiator support frame, then refer to paragraph 4 and remove the axle assembly. Attach a hoist or other supporting device to the front support, then unbolt and separate the front support from the front of the engine. Be careful not to lose shims which may be located between front support casting and front of engine.

Reattach front support to engine of all models by reversing the removal procedure, but omitting any shims originally installed between support and engine. On 340-360 models, tighten the retaining screws to 240-320 N·m (177-236 ft.-lbs.) torque. Measure any gap (A—Fig. 12) between the top spacer tube on right side and front support with a feeler gauge. Loosen retaining screws and install shims equal to the measured gap plus 0.13 mm (0.005 inch).

Fig. 13—Use a feeler gauge to measure gap (A) between the lower ears of casting and spacer tubes for 399 models.

Tighten screws retaining front support of 399 models to 230-255 N·m (170-190 ft.-lbs.) torque. Use a feeler gauge to measure any gap between the lower ears of front support casting and spacer tubes of 399 models as shown in Fig. 13. Loosen retaining screws

and install shims equal to the measured gap plus 0.13 mm (0.005 inch).

On all models, complete assembly by reversing the removal procedure.

FRONT-WHEEL DRIVE

7. A mechanical front-wheel drive is available on these models. There are some differences between the Front-Wheel Drive Systems used on these models that will be referred to in the servicing instructions which follow.

The front drive is engaged by an electric solenoid/hydraulic valve which directs oil pressure to move a dog clutch or by a dog clutch which is moved manually by shift fork by way of mechanical linkage. The transfer gearbox is attached to the left side of the range gearbox of models with twelve-speed shuttle transmissions or located between the rear of the transmission housing and the front of the rear axle housing of other models. On all models, a drive shaft with two "U" joints connects the transfer gearbox to front axle.

TRACK ROD AND TOE-IN

All Four-Wheel-Drive Models

8. All models are equipped with hydrostatic steering. A single track rod connects the left and right steering arms which are attached to the steering spindles. The track rod assures that both left and right wheels turn in unison, and the distance between ends of track rod establishes front wheel toe-in.

Ends of track rod are automotive type and should be renewed if wear is excessive. The procedure for removing and installing ends is self-evident. Recom-

mended toe-in is 0 for all models. Toe-in should be measured between the wheel rims on center line of axle, parallel to ground. Rotate wheels and remeasure to be sure that wheels are not bent giving incorrect reading.

Distance between front wheels is adjustable to different widths by relocating the wheel on the center disc or by reversing the wheels. If wheels are reversed, they must be moved to opposite sides of tractor to maintain correct tire tread direction.

To adjust toe-in, loosen the locknuts at each end of the track (tie) rod, then turn the tie rod tube to set the toe-in. Tighten locknuts at each end when adjustment is correct. Nut retaining ball-joint of track rod end in the steering arm should be tightened to 78-86 N·m (58-63 ft.-lbs.) torque for 340-360 models, 108-118 N·m (80-87 ft.-lbs.) torque for 399 models.

DRIVE SHAFT

All Models So Equipped

9. REMOVE AND REINSTALL. To remove drive shaft, first loosen clamps (1 and 2—Fig. 14), then slide guard (3) into the center tube. Remove spring clip (6) from guard at front. Remove bolts (4 and 5) from ends, then remove the shield and drive shaft assembly. Unscrew seal retaining ring from the sliding coupling, remove sliding coupling from rear end

Fig. 14—View showing both ends of drive shaft and shield. Front (axle) end of shield is retained by clip (6).

1. Clamp
2. Clamp
3. Guard
4. Retaining bolts

5. Retaining bolts
6. Clip
7. Clamp

of drive shaft, then remove drive shaft from guard tube.

When reassembling, grease splines of sliding coupling and both universal joints. Insert the long section of drive shaft into guard, then assemble sliding coupling at rear, making sure that arrows on drive shaft and sliding coupling are aligned. Coat threads of bolts (4 and 5) with "Loctite 270," then attach drive

shaft flanges with bolts (4 and 5) tightened to 55-75 N•m (40-55 ft.-lbs.) torque. Install clip (6) at front (axle end), making sure that drain hole in guard is toward bottom. Distance from step in axle and front of guard should be approximately 90 mm (3.5 inches).

FRONT DRIVE AXLE

All Models So Equipped

10. R&R ASSEMBLY. First remove the drive shaft and shield as outlined in paragraph 9. Raise front of tractor in such a way that it will not interfere with the removal of the axle. Remove front wheels and weights, then support the axle with a suitable jack or special safety stand to prevent tipping while permitting the axle to be lowered and moved safely. Disconnect hoses from the steering cylinder and cover openings to prevent the entry of dirt. Remove retaining pin or screw from the front pivot, then use a suitable puller to withdraw axle pivot pin. Lower axle until free, then carefully roll axle away.

Reinstall front drive axle by reversing the removal procedure. A spacer is located at the rear of the axle pivot and sufficient thickness of shims should be located in front of axle to reduce end play to 0.05-0.25 mm (0.002-0.010 inch). Grease pivot after correct thickness of shims and pivot pin retaining screw or pin have been installed.

PLANETARY ASSEMBLY

All Models So Equipped

11. R&R AND OVERHAUL. Refer to Fig. 15. Either planetary assembly can be serviced without

Fig. 15—Cross section of typical front drive axle.

removing the steering knuckle housing (26—Fig. 16). However, if seal (24) or bearings (20 and 23) are to be serviced, hub (22) and knuckle housing (26) must first be removed as outlined in paragraph 12. Support front axle housing and remove front wheel. Rotate hub (22) until drain plug (1), is down, then remove plug and drain oil from hub assembly. Remove screws (2) and lift off planetary carrier (3). Remove snap ring (12), then bump shafts (10) from housing bores. Keep each planet assembly, consisting of shaft (10), gear (7), bearings (8) and thrust washers (6), separate so that they can be reinstalled as a set. The sun gear (13) can be removed after removing snap ring (9).

Clean and inspect all parts for excessive wear or other damage and renew as necessary.

Install snap rings (11), bearings (8), planetary gears (7) and thrust washers (6), then secure with snap ring (12). Opening in snap ring (12) must b positioned in center of one of the three blocks, not in an open area. Make sure thrust plate (5) is in place. Clean mating surfaces of wheel hub (22) and planet carrier (3), then coat face with "Loctite 515 Instant Gasket" and install carrier assembly. Coat screws (2) lightly with oil before installing, then tighten to 96-118 N•m (71-87 ft.-lbs.) torque.

Turn wheel hub (22) until hole for plug (1) is horizontal, fill hub and planetary to the level of opening for plug (1) with an approved oil, then install and tighten plug. Recommended oil for use in front drive axle is Massey-Ferguson Super 500 multi-use 10W-

Fig.16—Exploded view of front-wheel-drive steering knuckle and planetary assembly.

1. Drain plug	12. Snap ring	31. "O" ring	
2. Screws	13. Sun gear	22. Wheel hub	31. "O" ring
3. Planet carrier	14. Ring gear	23. Bearing cup and cone	32. Bearings
4. "O" ring	15. Screws	24. Seal	33. Pivot pin
5. Thrust pin	16. Snap ring	25. Track rod and rod end	35. Upper retainer
6. Thrust washer	17. Ring gear hub	26. Steering knuckle	36. Screw
7. Planet gear	18. Pins	27. Axle center housing	37. Bushing
8. Rollers (50/gear)	19. Shims (0.10, 0.15, 0.30,	28. Bushing	38. Seal
9. Snap ring	0.50, 0.70, 1.00 mm)	29. Lower retainer	39. Axle shaft assy.
10. Planet gear shaft	20. Bearing cup and cone	30. Shims (0.10, 0.15, 0.20,	40. Seal
11. Snap ring	21. Steering cylinder	0.50, 1.00 mm)	41. Bushing

30 oil, Massey-Ferguson Permatran or equivalent. Install front wheel and tighten disc to hub nuts to 270 N•m (200 ft.-lbs.) torque. Disc to wheel rim nuts should be tightened to 190 N•m (140 ft.-lbs.) torque.

STEERING KNUCKLE HOUSING AND WHEEL HUB

All Models So Equipped

12. R&R AND OVERHAUL. To remove either steering knuckle housing, first remove planetary as outlined in paragraph 11. Remove snap ring (9—Fig. 16) and sun gear (13). Disconnect steering cylinder (21) from right side steering knuckle arm, and disconnect track rod (25) from steering knuckle arms on both sides. Unbolt and remove upper and lower retainers (29 and 35) and shims (30). Measure and note thickness and number of shims (30) under each retainer for aid in reassembly. Carefully remove the bearing cups for upper and lower bearings (32). Use a suitable puller such as MF 451A or equivalent to pull upper and lower pivot pins (33). Carefully remove steering knuckle housing (26) and wheel hub assembly (22) from axle center housing (27), leaving axle shafts (39) with center housing.

Axle shaft and double "U" joint assembly (39) may be withdrawn from center housing for inspection or repair. If renewal is required, oil seal (38) and bushing (37) can be removed from knuckle housing (26) and oil seal (40) and bushing (41) can be removed from axle housing. Bushings (37 and 41) should be pressed into position with external groove toward top and internal arrow-shaped grooves pointing toward inside of oil-filled housing (away from seal). Be careful not to damage seals (38 and 40) when installing axle and knuckle housing.

To disassemble the wheel hub and steering knuckle, remove the six cap screws (15). Attach an internal bearing puller as shown in Fig. 17 and pull ring gear hub (17—Fig. 16) and ring gear (14) from pins (18). Be careful not to lose or damage shims (19). If puller is not available, it may be possible to use a punch through hole for plug (1) to carefully drive hub and ring gear from pins. Solid pins are used on later models in place of the three split pins (18). Remove pins only if necessary. Bump hub (22) and bearings (20 and 23) from steering knuckle (26). Remove cups for bearings (20 and 23) and seal (24) from hub. If necessary, remove inner bearing cone from knuckle housing.

Clean and inspect all parts for excessive wear or other damage and renew as necessary.

The steering knuckle and wheel hub can be assembled, then attached to the axle center member; however, the assembly is heavy and difficult to handle. End play of the steering knuckle bearings (32) must

be measured and adjusted to provide 0.2 mm (0.008 inch) preload. End play measurement and adjustment is easier if the steering knuckle (26) is attached to the axle center member (27) before wheel hub and ring gear are assembled to the knuckle.

Be sure bushings (37 and 41) and seals (38 and 40) are installed, carefully locate the drive shaft and "U" joint assembly (39) in axle center member, then slide steering knuckle (26) over outer end of axle shaft. Lubricate "O" rings (31) and install in grooves of pivot pins (33) then press pins through steering knuckle bores and into bores in axle center member. Caps (29 and 35) and retaining screws (36) can be used to press pins (33) into bores if opposing screws (36) are carefully and evenly tightened. Be sure pins are fully seated in axle center member, remove caps (29 and 35) and install inner cones for both bearings (32) on pins. Install outer race for the lower bearing, install lower bearing cap and tighten the four retaining screws (36) to 96-118 N•m (71-87 ft.-lbs.) torque. Install the outer race for upper bearing, install upper cap (35) without shims (30) and tighten retaining screws to 96-118 N•m (71-87 ft.-lbs.) torque. Use a dial indicator and measure end play of the steering knuckle (26). Be sure that pins (33) and bearings (32) are fully seated. Without any shims (30), steering knuckle should have some end play. Remove upper cap (35), install shims (30) equal to the measured end play plus the desired bearing preload of 0.2 mm (0.008 inch), then reinstall upper cap (35). Tighten

Fig. 17—View showing bearing puller attached for pulling the ring gear from steering knuckle. Refer to text.

Fig. 18—Measure standout of inner bearing race as shown when selecting shims (19—Fig. 16). Refer to text.

cap retaining screws to 96-118 N·m (71-87 ft.-lbs.) torque. Shims (30) can normally be added to the upper cap (35), but if a large number of shims are necessary, shims may be divided between the upper and lower caps to center the bearings (41 and 37) for axle (39).

Observe the following when reassembling wheel hub (22) to the steering knuckle (26). Drive the cups for bearings (20 and 23) and the oil seal (24) into hub (22). Lubricate bearings, bushings and seals liberally.

Fig. 19—Measure standout of ring gear hub as shown when selecting shims (19—Fig. 16). Refer to text.

If removed, install new cone for inner bearing (23) on knuckle housing (26). Install hub (22) and cone for outer bearing (20) onto knuckle housing. It may be necessary to heat bearing cones before installing. Install two screws with washers in opposing holes as shown in Fig. 18 to hold bearing (20—Fig. 16) in position while measuring standout of bearing race. Be sure that bearings are fully seated, then measure distance from end of steering knuckle hub and outer surface of outer bearing inner race. Measure standout of ring gear hub from the bearing contact surface to the steering knuckle hub contacting surface as shown in Fig. 19. Subtract measurement of ring gear hub (Fig. 19) from the bearing standout (Fig. 18), then add the desired bearing preload, 0.05 mm (0.002 inch) to determine the correct thickness of shims to install.

An example is:

Measurement shown in Fig. 18	8.10 mm	0.319 inch
Measurement shown in Fig. 19	− 6.20 mm	− 0.244 inch
	1.90 mm	0.075 inch
Desired bearing preload	+ 0.05 mm	+ 0.002 inch
Thickness of shims to install	1.95 mm	0.077 inch

Please note that measurements above are only examples to indicate the method of determining the correct thickness of shims to install. Selecting and installing the proper thickness of shims (19—Fig. 16) for each combination of parts will provide bearings (20 and 23) with correct, 0.05 mm (0.002 inch), preload.

After selecting the correct thickness of shims (19), install pins (18) if removed. If pins (18) are split type, install largest pin first, followed by medium size with gap located 180° from gap of the largest pin. Install smallest of the three pins with gap 180° from the middle size pin or nearly aligned with groove of the largest pin. On all models, remove screws (S—Fig. 18) and washers used while measuring. Install wheel hub (22), steering knuckle (26), ring gear (14), hub (17), shims (19) and related parts. Coat threads of screws (15) with "Loctite 270" and tighten evenly to 112-123 N·m (83-91 ft.-lbs.) torque.

Make sure thrust plate (5) is in place. Clean mating surfaces of wheel hub (22) and planet carrier (3), then coat face with "Loctite 515 Instant Gasket" and carrier assembly. Coat screws (2) lightly with oil before installing, then tighten to 96-118 N·m (71-87 ft.-lbs.) torque. Nut retaining ball-joint of track rod end in the steering arm should be tightened to 78-86 N·m (58-63 ft.-lbs.) torque for 340-360 models, 108-118 N·m (80-87 ft.-lbs.) torque for 399 models. Nut retaining ball-joint of steering cylinder end in the steering arm should be tightened to 98-108 N·m (72-80 ft.-lbs.)

Fig. 20—View of removed differential assembly, showing marks (M) on one of the caps (C) and housing. Mark the other cap so that caps can be quickly identified for assembly to the correct side and in the correct position.

torque for 340-360 models, 177-196 N·m (130-145 ft.-lbs.) torque for 399 models.

Turn wheel hub (22) until hole for plug (10) is horizontal, fill hub and planetary to the level of opening for plug (1) with an approved oil, then install and tighten plug. Recommended oil for use in front drive axle is Massey-Ferguson Super 500 multi-use 10W-30 oil, Massey-Ferguson Permatran or equivalent. Install front wheel and tighten disc to hub nuts to 270 N·m (200 ft.-lbs.) torque. Disc to wheel rim nuts should be tightened to 190 N·m (140 ft.-lbs.) torque.

DIFFERENTIAL

All Except Models With "Hydralock"

13. REMOVE AND REINSTALL. The pinion shaft oil seal can be removed and a new seal installed, without removing the complete differential assembly. Mark the position of the pinion nut before removing, so that nut can be reinstalled without disturbing pinion bearing preload. Other service to the differential assembly requires removal of the unit.

To remove the differential assembly from front axle, refer to paragraph 9 and remove the drive shaft, then paragraph 12 and remove steering knuckles and axle shaft assemblies (39—Fig. 16) from both sides. Unbolt and remove the differential housing from the axle center housing. Some mechanics prefer to remove the front axle from tractor as outlined in paragraph 10 before removing steering knuckles and axle shafts.

Before reinstalling differential, clean mating surface of differential housing (1—Fig. 21) and axle center housing (27—Fig. 16), then coat mating surface with "Loctite 515 Instant Gasket." Lubricate screws attaching differential to axle housing, install differential assembly and tighten retaining screws to 96-118 N·m (71-87 ft.-lbs.) torque. Install drive shaft coupling on pinion shaft, coat threads of center retaining screw with "Loctite 270," then tighten center screw to 60 N·m (44 ft.-lbs.) torque. Complete assembly by reversing the removal procedure. Refer to paragraph 9 for installing drive shaft.

14. OVERHAUL. Before disassembling, mark both bearing caps and housing as shown at (M—Fig. 20) to facilitate alignment when reassembling. Straighten tabs of lock plate (13—Fig. 21), then remove cap screw (12) and lock plate (13). Loosen, but do not remove, the four screws attaching caps (C), then unscrew adjusting rings (14). Remove both bearing caps and lift differential, bearing cups and adjusting rings from housing.

NOTE: Some models are equipped with threaded adjusting rings (14—Fig. 20 and Fig. 21) on both sides, while others are equipped with spacer and shims (24 and 25—Fig. 22) on one side.

To remove the bevel pinion (10—Fig. 21), unstake nut (2), then remove nut and spacer (3). Push pinion out toward inside. Cone of bearing (7) will slide from shaft as pinion is removed.

Mark differential case (16 or 27) and ring gear (11) to facilitate alignment when reassembling. On models with "Autolock" differential, refer to Fig. 24 and install washers of proper size and through-bolt as shown to hold the differential together. On all models, remove screws (17—Fig. 21), then remove ring gear (11) and separate case (16). Note that two of the screws (17) are close fitting dowel screws and must be installed in same location when reassembling. Individual parts of the "Autolock" differential (Fig. 23) are not available and parts must be assembled in original location. Regardless of type, assembly of original parts is easier if parts are kept together for each side and not mixed.

Clean and inspect all parts for wear or other damage. Lubricate all parts while assembling. Align previously affixed marks on case (16 or 27—Fig. 21) and ring gear (11) while assembling. Coat threads of screws (17) with "Loctite 270," install and tighten screws to 79-87 N·m (58-64 ft.-lbs.) torque. Press bearing cones (15) onto case (16) until seated. Remove through-bolt and washers (Fig. 24) from "Autolock" models.

On all models, thickness of shims (9—Fig. 21) should be selected to change mesh position of bevel pinion (10) if gear set (10 and 11), pinion bearings (7 and 8) and/or differential housing (1) is renewed or if previous mesh position is questioned. The number stamped on the end of the pinion gear is the deviation from the standard distance of 118 mm or 113 mm (early axles with two adjuster rings for differential). Determine machined (cone) position by adding the stamped number (if marked with "+") or subtracting the stamped number (if marked with "–") from the standard distance. Install cup for bearing (8) in housing (1), then position cone for bearing (8) in cup as shown in Fig. 25. Measure distance (A—Fig. 25) from center line of the differential, as measured across the bearing saddle, to the face of inner race for pinion bearing (8). Subtract the measured distance (A—Fig. 25) from the machined cone distance (determined by adding or subtracting the number stamped on end of pinion to or from the standard dimension). The result is the thickness of shims (9—Fig. 21) to add under the bearing cup (8).

An example is:

Standard measurement	118.00 mm
Dimension stamped on pinion	+ 0.20
Machined distance	118.20 mm
Measurement (A—Fig. 25)	119.50 mm
Machined distance	– 118.20 mm
Required shim thickness	1.30 mm

Please note that measurements above are only examples to indicate the method of determining the correct thickness of shims to install. Selecting and installing the proper thickness of shims (9—Fig. 21) for each combination of parts will provide gear set with proper mesh position. Also, note standard measurement (cone point) of some models is 113 mm.

Fig. 21—Exploded view of standard differential assembly typical of the type used on many models. Gears (10 and 11) are available only as matched sets. "Autolock" differential unit (26) is shown in Fig. 23.

1. Housing
2. Nut
3. Spacer
4. Seal
5. Seal
6. "O" ring
7. Bearing cup & cone
8. Bearing cup & cone
9. Shim
10. Bevel pinion
11. Ring gear
12. Screw
13. Locking clip
14. Adjusting ring
15. Bearing cup & cone
16. Differential case
17. Screws
18. Thrust washers
19. Pinion gears
20. Pinion shaft
22. Side gears
23. Thrust washer
24. Shims
25. Spacer
26. "Autolock" differential
27. "Autolock" differential case

Fig. 22—Cross sectional view of differential of the type shown in Fig. 21. Refer to Fig. 21 for legend.

Remove cup for bearing (8), install correct thickness of shims (9), then press cups for both bearings (7 and 8) into housing bores. Shims are available in various thicknesses from 0.10 mm to 1.00 mm. Install pinion (10), cone for bearing (7), "O" ring (6), seals (5 and 4) and spacer (3). Lubricate all parts liberally while assembling. Install a new special nut (2), tightening nut until 2.0-2.5 N·m (18-22 in.-lbs.) torque is

required to turn pinion in the bearings. When correct rotational force is obtained, stake nut to shaft spline.

Reinstall differential and ring gear in housing (1—Fig. 21) by reversing removal procedure. Adjust gear backlash and carrier bearings as follows.

Models with one adjuster ring (14—Fig. 20 and Fig. 21) are fitted with shims (24—Fig. 21) located on right side between spacer (25) and right side carrier bearing. The shims (24) are used to adjust backlash between the ring gear (11) and pinion (10). When assembling, install shims originally installed, install bearing caps (C) and tighten screws retaining caps to 136-150 N·m (100-110 ft.-lbs.) torque. Tighten adjuster ring (14) on left side until all clearance is just removed from carrier bearings (15), then tighten adjuster ring (14) three additional notches to provide correct bearing preload and align notch so that lock plate (13) can be installed. Do not yet install lock plate. Mount a dial indicator as shown in Fig. 26 and measure backlash between ring gear teeth and pinion gear teeth. Correct backlash is 0.10-0.25 mm (0.004-0.010 inch) when measured as shown. Reduce thickness of shims (24—Fig. 21) if backlash is excessive; increase thickness of shims if backlash is too tight.

On models with two adjuster rings (14—Fig. 20 and Fig. 21), one on each side, the setting of adjuster rings adjusts backlash of the bevel gear teeth and preload of carrier bearings. When assembling, install bearing caps (C) and tighten screws retaining caps to 136-150 N·m (100-110 ft.-lbs.) torque. Tighten both adjuster rings (14) until all clearance is just removed from carrier bearings (15), but be sure that some backlash exists between gear teeth. Tighten adjuster ring (14) on right side three additional notches to provide correct bearing preload and align notches so that lock plates (13) can be installed. Do not yet install lock plate. Mount a dial indicator as shown in Fig. 26 and measure backlash between ring gear teeth and pinion gear teeth. Correct backlash is 0.10-0.25 mm (0.004-0.010 inch) when measured as shown. Loosen adjust-

Fig. 23—Exploded view of "Autolock" differential. Washers and through-bolt shown in Fig. 24 must be used to compress springs (3) and hold assembly together while removing and installing.

1. Side gears
2. Spring retainers
3. Springs
4. Driven clutch assy.
5. Center cam
6. Central driver & center cam assy.
7. Snap ring
8. Holdout rings
9. Driven clutch

Fig. 24—Washers and a through bolt should be used to hold "Autolock" differential assembly together while removing and installing. Diameter of washers should be small enough to pass through opening (A) in case (27—Fig. 21), but large enough to press against splines for axles in side gears (1—Fig. 23).

Fig. 25—Measure distance (A) from center line of differential, across bearing saddle, to the face of inner race for bearing (8).

ing ring (14—Fig. 21) on left (ring gear) side, then turn adjuster on right side to change backlash. Readjust bearings by tightening adjuster ring on ring gear (left) side until all clearance is just removed from carrier bearings (15), then tighten left side adjuster ring three additional notches to provide correct bearing preload and align notches so that lock plates (13) can be installed.

Fig. 26—View of typical differential assembly with dial indicator (D) mounted for checking gear backlash. Refer to text for procedure.

On all models, recheck backlash after making any changes. With gear backlash, pinion bearing preload and carrier bearing preload correctly set, rotational torque of pinion shaft should be 3.2-3.3 N·m (28.32-29.21 in.-lbs.) torque. After it is determined that backlash and bearings are correctly set, carefully remove one of the four screws retaining caps (C), coat threads with "Loctite 270," then reinstall and tighten the screw to 136-150 N·m (100-110 ft.-lbs.) torque. Remove each of the remaining cap retaining screws one at a time, coat with "Loctite" and reinstall at correct torque. Coat threads of screws (12) with "Loctite 270," then install locking tabs (13) and retaining screws (12). Tighten screws (12) to 16-26 N·m (12-19 ft.-lbs.) torque.

Models With "Hydralock"

15. REMOVE AND REINSTALL. The pinion shaft oil seal can be removed and a new seal installed, without removing the complete differential assembly. Mark the position of the pinion nut before removing, so that nut can be reinstalled without disturbing pinion bearing preload. Other service to the differential assembly requires removal of the unit.

To remove the differential assembly, refer to paragraph 9 and remove the drive shaft, then paragraph 12 and remove steering knuckles and axle shaft assemblies (39—Fig. 16) from both sides. Detach hydraulic hose and cover openings to prevent entrance of dirt. Unbolt and remove the differential housing from the axle center housing. Some mechanics prefer to remove the front axle from tractor as outlined in paragraph 10 before removing steering knuckles and axle shafts.

Before reinstalling differential, clean mating surface of differential housing (1—Fig. 27) and axle center housing (27—Fig. 16), then coat mating surface with "Loctite 515 Instant Gasket." Lubricate screws attaching differential to axle housing, install differential assembly and tighten retaining screws to 96-118 N·m (71-87 ft.-lbs.) torque. Install drive shaft coupling on pinion shaft, coat threads of center retaining screw with "Loctite 270," then tighten center

screw to 60 N·m (44 ft.-lbs.) torque. Complete assembly by reversing the removal procedure. Refer to paragraph 9 for installing drive shaft.

16. OVERHAUL. Before disassembling, mark both bearing caps and housing as shown at (M—Fig. 20) to facilitate alignment when reassembling. Straighten tabs of lock plate (13—Fig. 27), then remove cap screw (12) and lock plate (13). Loosen, but do not remove, the four screws attaching caps (C), then unscrew adjusting ring (14). Carefully remove both bearing caps, noting that the cap nearest the locking assembly retains a spring under compression. The spacer (25), spring retaining plate (28) and spring (29) should remain in the housing. Lift differential and carrier bearing cones from housing. Keep

Fig.27—Exploded view of "Hydralock" differential assembly. Gears (10 and 11) are available only as matched sets. Drawing shows assembly inverted. Ring gear (11) is on right side of pinion (10).

1. Housing	10. Bevel pinion	18. Thrust washers
2. Nut	11. Ring gear	19. Pinion gears
3. Spacer	12. Screw	20. Pinion shaft
4. Seal	13. Locking clip	22. Side gears
5. Seal	14. Adjusting ring	23. Thrust washer
6. "O" ring	15. Bearing cup & cone	24. Shims
7. Bearing cup & cone	16. Differential case	25. Spacer
8. Bearing cup & cone	17. Screws	28. Spring guide plate
9. Shim		

29. Spring
30. Sleeve
31. Thrust washer
32. Piston
33. Dowel
34. Thrust washer
35. Snap ring
36. "O" ring
37. "O" ring

Fig. 28—Cross section of "Hydralock" unit with spring compressor (SC) in position. Fluid port is identified by (FP). Refer to Fig. 27 for remainder of legend.

cups for carrier bearings (15) with the respective cones.

Install Massey-Ferguson MF471 or equivalent spring compressor (SC—Fig. 28) to compress spring (29—Fig. 27), then remove shims (24) and spacer (25). Bump the "Hydralock" unit (Fig. 28) toward center and remove from housing. "O" rings (36 and 37) are located in housing.

To remove the bevel pinion (10—Fig. 27), unstake nut (2), then remove nut and spacer (3). Push pinion out toward inside. Cone of bearing (7) will slide from shaft as pinion is removed.

Mark differential case (16) and ring gear (11) to facilitate alignment when reassembling. Remove screws (17), then remove ring gear (11) and separate case (16). Assembly is easier if parts are kept together for each side and not mixed.

Clean and inspect all parts for wear or other damage. Lubricate all parts while assembling. Align previously affixed marks on case (16) and ring gear (11) while assembling. Coat threads of screws (17) with "Loctite 270," install and tighten screws to 79-87 N•m (58-64 ft.-lbs.) torque. Press bearing cones (15) onto case (16) until seated.

Thickness of shims (9—Fig. 27) should be selected to change mesh position of bevel pinion (10) if gear set (10 and 11), pinion bearings (7 and 8) and/or

differential housing (1) is renewed or if previous mesh position is questioned. The number stamped on the end of the pinion gear is the deviation from the standard distance of 118 mm or 113 mm (early axle with two adjuster rings for differential). Determine machined (cone) position by adding the stamped number (if marked with "+") or subtracting the stamped number (if marked with "–") from the standard distance. Install cup for bearing (8) in housing (1), then position cone for bearing (8) in cup as shown in Fig. 25. Measure distance (A—Fig. 25) from center line of the differential, as measured across the bearing saddle, to the face of inner race for pinion bearing (8). Subtract the measured distance (A—Fig. 25) from the machined cone distance (determined by adding or subtracting the number stamped on end of pinion to or from the standard dimension). The result is the thickness of shims (9—Fig. 27) to add under the bearing cup (8).

An example is:

Standard measurement	118.00 mm
Dimension stamped on pinion	+ 0.20
Machined distance	118.20 mm
Measurement (A—Fig. 25)	119.50 mm
Machined distance	– 118.20 mm
Required shim thickness	1.30 mm

Please note that measurements above are only examples to indicate the method of determining the correct thickness of shims to install. Selecting and installing the proper thickness of shims (9—Fig. 27) for each combination of parts will provide gear set with proper mesh position. Also, note standard measurement (cone point) of some models is 113 mm.

Remove cup for bearing (8), install correct thickness of shims (9), then press cups for both bearings (7 and 8) into housing bores. Shims are available in various thicknesses from 0.10 mm to 1.00 mm. Install pinion (10), cone for bearing (7), "O" ring (6), seals (5 and 4) and spacer (3). Lubricate all parts liberally while assembling. Install special nut (2), tightening nut until 2.0-2.5 N•m (18-22 in.-lbs.) torque is required to turn pinion in the bearings. When correct rotational force is obtained, stake nut to shaft spline.

Install and lubricate "O" rings (36 and 37—Fig. 27) in housing (1), then slide piston (32) into position making sure that pin (33) is in correct location. Install sleeve (30), thrust washers (31 and 34) and snap ring (35). Position spring (29) and retainer (28), then install spring compressor (SC—Fig. 28) so that spacer (25—Fig. 27) can be installed in housing groove.

Install spacer (25) and originally installed shims (24), then remove the spring compressor (SC—Fig. 28). Reinstall differential and ring gear in housing (1), then adjust gear backlash and carrier bearings as follows.

One adjuster ring (14) is fitted and shims (24—Fig. 21) are located on right side between spacer (25) and right side carrier bearing. The shims (24) are used to adjust backlash between the ring gear (11) and pinion (10). When assembling, install shims originally installed, install bearing caps (C) and tighten screws retaining caps lightly. Use a pry bar to move the differential against the pressure of spring (29) and to remove all play in bearings (15), while tightening adjuster ring (14) to just remove all play. Tighten adjuster ring (14) three additional notches to provide correct bearing preload and align notch so that lock plate (13) can be installed after all clearance is just removed from carrier bearings (15). Do not yet install lock plate. Mount a dial indicator as shown in Fig. 26 and measure backlash between ring gear teeth and pinion gear teeth. Correct backlash is 0.10-0.25 mm (0.004-0.010 inch) when measured as shown. Reduce thickness of shims (24) if backlash is excessive; increase thickness of shims if backlash is too tight. Recheck backlash after making any changes. After determining that backlash and bearings are correctly set, carefully remove one of the four screws retaining caps (C), coat threads with "Loctite 270," then reinstall and tighten the screw to 136-150 N•m (100-110 ft.-lbs.) torque. Remove each of the remaining cap retaining screws one at a time, coat with "Loctite 270" and reinstall at correct torque. Coat threads of screw (12) with "Loctite 270," then install locking tab (13) and retaining screw (12). Tighten screw (12) to 16-26 N•m (12-19 ft.-lbs.) torque.

TRACTOR SPLIT

Four-Wheel-Drive Models Without Cab

17. To separate tractors (without cab) between the rear of the transfer gearbox and the front of the rear axle center housing, proceed as follows:

Set parking brake and block the rear wheels to prevent rolling. Disconnect battery ground cable and drain all transmission fluid. Refer to paragraph 9 and remove the front drive shaft. Remove lower (gearbox) cover, both foot steps and battery boxes. Detach foot throttle, four-wheel-drive indicator switch and four-wheel-drive selector lever. Disconnect hydraulic oil supply pipe located on right side of tractor and, if so equipped, detach hydraulic pipes from left side. Disconnect brake lines and drain system. Disconnect wiring harness at the connectors below console on right side and disconnect wires from safety start switch. Remove the hydraulic suction filter assembly. Place hardwood wedges between the front axle and axle support casting on both sides to prevent tipping around axle pivot when tractor is separated. Support rear of tractor under axle center housing and front of tractor under transmission housing. Remove screws attaching transfer case to the rear axle center hous-

ing, then separate the tractor halves. Either the front can be moved forward or the rear can be moved to the rear, but it is important to support both halves safely and securely.

If the split pin is removed from the shear tube, install pin in center of the five holes. Install shear tube on pinion. Install at least two dowel studs to facilitate alignment, position new gasket on dowel studs and position rear drive shaft into high/low gear coupler. Move the high/low shifter to "low" and shift main transmission to 3rd. Move tractor together, turning engine to align splines of shear tube. Make sure that flanges are tight against each other before installing retaining screws. Install and tighten one screw on each side, then check end float of the rear drive shaft. Push the shear tube toward rear as shown at inset in Fig. 29, which will compress spring, then measure clearance between the shear tube and locking collar. Correct clearance is 0.4-2.5 mm (0.015-0.100 inch) and can be changed by separating tractor and moving the split ring to another hole in the shear tube. Tighten all screws retaining the transfer case to the rear axle center housing to 112 N•m (83 ft.-lbs.) torque, beginning at top center and progressing in a clockwise direction (as viewed from tractor rear) two times around the mating flange. Remainder of assembly is reverse of disassembly.

Four-Wheel-Drive Models With Cabs

18. To separate tractors (with cab) between the rear of the transfer gearbox and the front of the rear axle center housing, proceed as follows:

Set parking brake, block the rear wheels to prevent rolling and drain all transmission fluid. Disconnect battery ground cable, then disconnect wiring harness at the connectors. Refer to paragraph 9 and remove the front drive shaft. Disconnect throttle stop cable and throttle cable from fuel injection pump and move cables out of the way so that cables will not catch when separating tractor. Disconnect and cap all hydraulic pipes running from front to rear, which would prevent separation of tractor halves. Mark (identify), disconnect and cap steering hoses, heater hoses and air conditioning lines (if so equipped), which would interfere. Place hardwood wedges between the front axle and axle support casting on both sides to prevent tipping around axle pivot when tractor is separated. Remove front floor mat and floor plates, disconnect clutch linkage and detach gearbox oil filler pipe. Unbolt and remove the gear shift cover complete with shift levers, then unbolt and remove the transmission cover. Disconnect wires at safety start switch and four-wheel-drive indicator switch. Remove the oil feed pipe and filter housing from the right side of transmission case. Disconnect four-wheel-drive selector lever, then check to make sure that all wiring, cables, linkages and hoses are free to permit the tractor to be separated. On model 340 tractors, disconnect the hydraulic pipe from inside the transmission housing which passes from the rear axle center housing to the clutch housing. On all models, support rear of tractor under axle center housing and at front corners of cab. Support front of tractor under transmission housing in such a way that front of tractor, engine, transmission and transfer case can be rolled forward (Fig. 29). Unbolt and remove both cab support brackets from front of cab. Remove screws attaching transfer case to the rear axle center housing, then move front of tractor forward. It is important to support both halves safely and securely to prevent damage and injury.

If the split pin is removed from the shear tube, install pin in center of the five holes. Install shear tube on pinion. Install at least two dowel studs to facilitate alignment, position new gasket on dowel studs and position rear drive shaft into high/low gear coupler. Move the high/low shifter to Low and shift main transmission to 3rd. Move front half of tractor toward rear turning engine to align splines of shear tube. Make sure that flanges are tight against each other before installing retaining screws. Install and tighten one screw on each side, then check end float of the rear drive shaft. Push the shear tube toward rear as shown at inset in Fig. 29, which will compress spring, then measure clearance between the shear tube and locking collar. Correct clearance is 0.4-2.5 mm (0.015-0.100 inch) and can be changed by separating tractor and moving the split ring to another hole in the shear tube. Tighten all screws retaining the transfer case to the rear axle center housing to

112 N·m (83 ft.-lbs.) torque, beginning at top center and progressing in a clockwise direction (as viewed from tractor rear) two times around the mating flange. Remainder of assembly is reverse of disassembly.

TRANSFER GEARBOX

Power for the front drive axle is taken from a gearcase and clutch (Fig. 36) attached to the left side of the range gearbox of models with the twelve-speed shuttle transmission. Power for the front axle of models equipped with other transmissions, is taken from a transfer gearbox (Fig. 30 or Fig. 32) located behind the tractor standard transmission. Refer to the appropriate following paragraphs for service.

Four-Wheel-Drive Models With All Transmissions Except Twelve-Speed Shuttle

19. R&R AND OVERHAUL. To remove the transfer gearbox, first split the tractor between the rear of the transfer gearbox and the front of the rear axle center housing as outlined in paragraph 17 or 18, then proceed as follows.

Position floor jack under the transfer gearbox, remove retaining screws and lower transfer gearcase from the tractor.

To disassemble the unit, refer to Fig. 30 and Fig. 31 or Fig. 32 and Fig. 34 and proceed as follows. Remove input gear (2), then unbolt and remove retainer (5).

Fig. 29—View of tractor with cab separated between the four-wheel-drive transfer case and the rear axle center housing. Special splitting stand and track is shown.

Drive input shaft (3) and bearing (7) from housing. Remove snap ring (16) from shaft (11), then unbolt and remove retainer (10). Drive gear and shaft (11) out toward rear and withdraw gear (13) from housing. Remove snap ring (24), then use a 12 mm screw in threaded end of shaft (21) to withdraw shaft from housing bore. Lift gear (19), bearings (18) and thrust washers (17 and 20) from housing. Remove the center screw from drive shaft coupling, then remove coupling and drive shaft guard.

On early models (before July 1991), unbolt and remove cover (40—Fig. 30). Remove screw (44) from fork (43), then remove shaft (46), shift fork (43) and shoes (42). Remove output shaft (37) and related parts. Be careful not to lose the three sets of detent balls and springs (38).

On later models (from July 1991), unbolt and remove cover (61—Fig. 32), then remove screw (60) from fork. Pull shaft (63) from housing (57) and remove fork (59). Unbolt housing (57) and withdraw output shaft (37) and related parts. Be careful not to lose the two detent balls and one spring (55) if actuator (54) is removed from shaft (37). Refer to Fig. 32 and Fig. 33.

Clean and inspect all parts and renew any showing excessive wear or other damage. When reassembling, renew all "O" rings, seals and gaskets. Reassemble by reversing disassembly procedure.

Fig. 30—Exploded view of four-wheel-drive transfer gearbox used on early models (before July 1991) without twelve-speed shuttle transmission. Refer to Fig. 33 for later type.

1. Snap ring	13. Gear	24. Snap ring	35. Output gear
2. Input gear	14. Snap ring	25. Pin	36. Engagement coupling
3. Input shaft	15. Ball bearing	26. Sleeve	37. Output shaft
4. Needle bearing	16. Snap ring	27. Retainer	38. Detent balls and springs (3)
5. Bearing retainer	17. Thrust washer	28. Seal	39. Ball bearing
6. Snap ring	18. Roller bearings (2)	29. Seal	40. Cover
7. Ball bearing	19. Intermediate gear	30. "O" ring	41. Bracket
8. Washer (340-360 models)	20. Thrust washer	31. Transfer gearbox housing	42. Shoes (2)
9. Snap ring	21. Intermediate shaft	32. Ball bearing	43. Shift fork
10. Retainer plate	22. "O" ring	33. Spacer	44. Set screw
11. Gear and shaft	23. Shim (0.1-0.2 mm)	34. Bushing	45. Seal
12. Roller bearing			46. Shift shaft

Fig. 31—Cross section of early style four-wheel-drive transfer gearbox shown exploded in Fig. 30.

Fig. 32—Exploded view of late style (from July 1991) four-wheel-drive transfer gearbox used on tractors without twelve-speed shuttle transmission.

1. Snap ring
2. Input gear
3. Input shaft
4. Roller bearing
5. Bearing retainer
6. Snap ring
7. Ball bearing
8. Washer (340-360 models)
9. Snap ring
10. Retainer plate
11. Gear and shaft
12. Roller bearing
13. Gear
14. Snap ring
15. Ball bearing
16. Snap ring
17. Thrust washer
18. Roller bearings (2)
19. Intermediate gear
20. Thrust washer
21. Intermediate shaft
22. "O" ring
23. Shim (0.1-0.2 mm)
24. Snap ring
25. Pin
26. Sleeve
27. Seal
29. Seal
30. "O" ring
31. Transfer gearbox housing
32. Ball bearing
33. Thrust washer
35. Output gear
47. Snap ring
48. Spring
49. Engagement coupling
50. Toggle (engagement) pins (3)
51. Abutment ring
52. Snap ring
53. Ball bearing
54. Actuator shaft
55. Detent spring and 2 balls
56. "O" ring
57. Shift housing
58. Pins (2)
59. Shift fork
60. Set screw
61. Cover
62. Seal
63. Shift shaft

Fig. 33—Cross section of late style four-wheel-drive transfer gearbox shown exploded in Fig. 32.

Fig. 34—View of four-wheel-drive transfer gearbox mounted on left side of tractor with twelve-speed shuttle transmission.

1. Driveshaft rear coupling
2. Wires to engagement solenoid   3. Hydraulic pressure line

drain oil from the transmission and detach rear of drive shaft from the coupling (1—Fig. 34). Disconnect hydraulic supply line (2) and wires (3). Support the gearbox, remove the retaining screws, then remove the unit from side of transmission case.

To disassemble the removed unit, first unscrew the solenoid valve (33—Fig. 35), remove the center screw (1) and remove the coupling (3). Remove screws (35) and cap (36), being careful not to lose shims (37), then push shaft (38) out of intermediate gear (41) and housing. Lift intermediate gear, bearings (39), snap rings (40) and thrust washer (42) from housing. Remove screws retaining cover (32), then remove cover, output shaft (20) and associated parts. Unbolt and remove cover (4), end cap (7), seals and associated parts being careful not to lose or damage shims (9). Press bearing cone (13) from shaft (20), then remove thrust washer (14) and gear (15). A steel tube with outside diameter of approximately 16 mm ($\frac{5}{8}$ inch), inside diameter of approximately 13 mm ($\frac{1}{2}$ inch) and approximately 100 mm (4 inches) long is necessary to remove and disassemble actuator piston (23). Use special tube and a press to push against washer (27) and compress spring (26) enough to remove snap ring (28). Remainder of disassembly will be self-evident. Refer also to cross section shown in Fig. 36.

Clean and inspect all parts and renew any showing excessive wear or other damage. When reassembling, renew all "O" rings, seals and gaskets. Reassemble as follows. Install gear (15—Fig. 35) and thrust washer (14) on output shaft (20), then press bearing cone (13) onto shaft against thrust washer and step on shaft. Assemble spring (16), coupler (17), abutment ring (18) and snap ring (19) on shaft, then press bearing

Reinstall transfer gearbox by reversing the removal procedure. Tighten transfer gearbox retaining screws to 112 N·m (83 ft.-lbs.) torque, beginning at top center and progressing in a clockwise direction (as viewed from tractor rear) two times around the mating flange. Coat threads of drive shaft retaining screws with "Loctite 270," then attach drive shaft flanges and tighten screws to 55-75 N·m (40-55 ft.-lbs.) torque. Remainder of assembly is reverse of disassembly.

Four-Wheel-Drive Models With Twelve-Speed Shuttle Transmission

20. R&R AND OVERHAUL. To remove the four-wheel-drive transfer gearbox for these models, first cone (29) onto output shaft (20). Install "O" rings (22 and 24) in grooves of actuator piston (23), lubricate piston and insert into bore of output shaft (20). Install toggle pins (21) between abutment ring (18) and coupler (17), making sure that pins enter the groove of actuator piston (23), then slide snap ring (19) into groove of output shaft (20). Install spring (26), sleeve (25) and washer (27), compress spring sufficiently so that snap ring (28) can be installed in groove of output shaft. Measure and adjust preload of bearings (13 and

29) before installing new seals (5 and 10). Assemble shaft assembly (13 through 29), rear cover (32) and front cover (7) using shims (9) originally installed, measure bearing preload and vary the thickness of shims (9) as required to establish correct preload of 0.07 mm (0.003 inch). When preload is correct, a slight drag should be felt when rotating the shaft by hand. There must be no free play in the bearings. Remove output shaft assembly to install intermediate gear and shaft assembly (38 through 42), measure

Fig. 35—Exploded view of transfer gearbox used on four-wheel-drive models with twelve-speed shuttle transmission.

1. Screw	12. "O" ring	23. Actuator piston	33. Solenoid valve
2. Washer	13. Tapered roller bearing	24. "O" ring	34. Housing
3. Driveshaft coupling	14. Thrust washer	25. Stop tube	35. Screw
4. Cover	15. Drive gear	26. Spring	36. End cap
5. Seal	16. Spring	27. Washer	37. Shims
6. Sleeve	17. Clutch coupler	28. Snap ring	38. Intermediate shaft
7. End cap	18. Abutment ring	29. Tapered roller	39. Tapered roller bearing
8. "O" ring	19. Snap ring	bearing	40. Snap rings
9. Shims	20. Output shaft	30. Insert	41. Intermediate gear
10. Oil seal	21. Toggle pins (3)	31. "O" ring	42. Thrust washer
11. Oil seal sleeve	22. "O" ring	32. End cover	43. Pin

Fig. 36—Cross sectional view of four-wheel-drive unit shown exploded in Fig. 35.

preload of bearings (39) and select correct thickness of shims (37). Install sufficient thickness of shims (37) to provide bearings (39) with 0-0.05 mm (0-0.002 inch) preload. When preload is correct, a slight drag should be felt when rotating the gear by hand. Reinstall shaft and bearing assembly (13 through 29), installing new "O" rings (8, 12 and 31) and seals (5 and 10). Remainder of reassembly is reverse of disas-

sembly. Degrease threads for screw (1), coat threads with "Loctite 270," then tighten to 113-137 N·m (83-101 ft.-lbs.) torque. Degrease mating surface of housing (34) and transmission, coat mating surfaces with "Loctite 515 Instant Gasket" and tighten retaining screws to 120-160 N·m (88-118 ft.-lbs.) torque. Tighten drive shaft coupling screws to 55-75 N·m (40-55 ft.-lbs.) torque.

STEERING SYSTEM

21. The hydrostatic steering system consists of a pump, steering valve assembly, and one or two steering cylinders. In normal operation, pressurized hydraulic fluid is supplied by the gear type hydraulic pump. However, in the event of hydraulic failure or engine stoppage, manual steering can be accomplished by the gerotor pump in the steering valve.

FILTER AND BLEEDING

All Models

22. Recommended steering fluid is Massey-Ferguson Super 500 multi-use 10W-30 oil or equivalent. On models with a separate steering system pump and reservoir, maintain fluid level at bottom of reservoir filler hole with tractor level. Fluid and filter should be changed if steering unit is overhauled or if fluid is suspected of contamination. The tandem (dual) pump, used on some models, is supplied with oil from the gearbox and rear axle center housing. A spin-on

filter is located near the pump to filter fluid returning from the steering control valve to the auxiliary section of the pump.

On models with separate pump, filter (3—Fig. 37) can be renewed as follows. Remove screw or nut (1) retaining reservoir (2), then pull reservoir from the pump unit and catch oil in a suitable container. Remove the filter assembly (3) and sealing ring (4). Clean the inside of the reservoir and outside of the pump. Install new sealing ring (4) and filter, then install reservoir, aligning breather at top before tightening screw or nut (1).

The hydrostatic steering system on all models is self-bleeding. When the unit has been disassembled, refill with new oil, then start engine and cycle the system several times by turning the steering wheel from lock to lock. Recheck fluid level and add fluid as required. Capacity of separate reservoir is approximately 0.7 L (1.2 pints). System is fully bled when no more air bubbles appear in reservoir as system is cycled.

Fig. 37—The separate power steering pump used on some three cylinder models is equipped with a filter as shown. Pump may be one of several different types, but reservoir and filter are serviced in a similar way.

TROUBLESHOOTING

All Models

23. Some of the problems which may occur during operation of power steering and their possible causes are as follows:

A. Steering wheel hard to turn. Could be caused by:
  a. Defective power steering pump.
  b. Leaking or missing recirculating ball.
  c. Mechanical parts of front steering system binding.
  d. Ball bearings in steering column damaged.
  e. Leaking steering cylinder.
  f. Control valve spool and sleeve binding.

B. Steering wheel turns on its own. Could be caused by:
  a. Leaf springs in control valve weak or broken.

C. Steering wheel does not return to neutral position. Could be caused by:

a. Control valve spool and sleeve jammed.
  b. Leakage between valve sleeve and housing.
  c. Dirt or metal chips between valve spool and sleeve.

D. Excessive steering wheel play. Could be caused by:
  a. Inner teeth of rotor or drive shaft teeth worn.
  b. Upper flange of drive shaft worn.
  c. Leaf springs in control valve weak or broken.

E. Steering wheel rotates at steering cylinder stops. Could be caused by:
  a. Excessive leakage in steering cylinder.
  b. Rotor and stator excessively worn.
  c. Excessive leakage between valve spool and sleeve
  d. Excessive leakage between sleeve and housing.

F. Steering wheel "kicks" violently. Could be caused by:
  a. Incorrect adjustment between drive shaft and rotor.

G. Steering wheel responds too slowly. Could be caused by:
  a. Excessive front end weight.
  b. Not enough oil.
  c. Steering control valve worn.

H. Tractor steers in wrong direction. Could be caused by:
  a. Hoses to steering cylinder incorrectly connected.
  b. Incorrect timing of drive shaft to rotor.

SYSTEM PRESSURE

Models With Separate Steering Pump

24. To check the system relief pressure, install a "T" fitting in pump pressure line at pump (Fig. 38), connect a 0-28,000 kPa (0-4000 psi) test gauge to the fitting and operate engine at 1200 rpm. Turn the front wheels to one extreme against lock, and observe gauge reading. System relief pressure should be 14,210 kPa (2030 psi) for three-cylinder models so equipped.

> **CAUTION: When checking system relief pressure, hold the steering wheel against lock only long enough to observe pressure indicated by gauge. Pump may be damaged if steering wheel is held in this position too long or if flow is otherwise stopped.**

The pressure relief valve is located in the steering control valve as shown at (34, 35 and 36—Fig. 52). Adjustment requires removal of the steering control valve as described in paragraph 34. Turning the relief valve spring seat (34) in, increases the pressure against spring (35) and increases system relief pressure. Spring seat plug (34) is accessible after removing plug (32).

Models With Tandem Pump

25. To check the system relief pressure, install a "T" fitting in pump pressure line at pump (Fig. 39), connect a 0-28,000 kPa (0-4000 psi) test gauge to the fitting and operate engine at 1200 rpm. Observe back pressure on gauge, then turn the front wheels to one extreme against lock, and observe gauge reading. Pressure when wheels are held against lock less the normal back pressure is the system relief pressure. System relief pressure should be 13,097-15,165 kPa (1900-2200 psi) for three-cylinder models, 15,855-17,923 kPa (2300-2600 psi) for six-cylinder models.

CAUTION: When checking system relief pressure, hold the steering wheel against lock only long enough to observe pressure indicated by gauge. Pump may be damaged if steering wheel is held in this position too long or if flow is otherwise stopped.

The pressure relief valve is located in the steering control valve as shown at (34, 35 and 36—Fig. 52). Adjustment requires removal of the steering control

Fig. 38—Steering system pressure can be checked on systems with separate pump as described in text after attaching pressure gauge as shown.

Fig. 39—Check steering system pressure with gauge attached as shown on models with tandem pump. System relief pressure is calculated by subtracting the circuit back pressure (pressure without turning wheels) from the higher pressure indicated on gauge with wheels held against one lock.

valve as described in paragraph 34. Turning the relief valve spring seat (34) in, increases the pressure against spring (35) and increases system relief pressure. Spring seat plug (34) is accessible after removing plug (32).

PUMP

All Models

26. A tandem (dual-section) pump is used on models equipped with spool valves and other auxiliary hydraulics. One section of the pump supplies pressure for the steering. Pressurized fluid not required for steering passes through a spin-on type filter. The second section of the pump supplies pressurized fluid to the auxiliary spool valves and independent pto clutch (if so equipped). Refer to paragraphs 29 and 30 for service to tandem hydraulic pump.

A separate steering pump containing its own reservoir is used on some three-cylinder models requiring hydraulic pressure for steering only. Refer to paragraphs 27 and 28 for service to the separate pump used on these models.

Models With Separate Pump

27. REMOVE AND REINSTALL. Clean pump and area around pump thoroughly, then disconnect lines from pump and allow fluid to drain. Cover all

Fig. 40—Exploded view of separate steering pressure pump used on some three-cylinder models. Unit shown is made by Sundstrand Hydraulic. Refer also to Fig. 41 for other separate pump.

1. Screw	
2. Reservoir	
3. Filter element	11. Cover and flange
4. "O" ring	12. Screw
5. Spring	14. Seals
6. Fill plug	15. Woodruff key
7. Gasket	16. Pump gears
8. Gasket	17. Bearings
9. Cover	18. Inner seals
	19. Outer seal

openings to prevent dirt from entering pump or lines, then unbolt and remove pump from the engine timing case.

Tighten pump retaining screws to 31-42 N·m (23-31 ft.-lbs.) torque. Refer to paragraph 22 for filling and bleeding.

28. OVERHAUL. Pumps from Sundstrand Hydraulic and Bendix (Hydraquip) have been used. Refer to the appropriate following paragraphs and Fig. 40 and Fig. 41.

Straighten tab of lock washer, remove nut, then use a puller to remove drive gear from pump shaft. Remove Woodruff key after gear is pulled from shaft. Remove screw or nut (1), then remove reservoir (2) and filter (3). Mark relative position of flange housing (11), body (10) and cover (9), then remove bolts (12). Pump gears (16) are available only as matched set. Check condition of bearing blocks (17), gears (16) and body (10) for wear or other damage.

Always use new filter and seals when assembling. Tighten pump body through-bolts (12) to 48-50 N·m (35-37 ft.-lbs.) torque and pump drive gear retaining nut to 40-45 N·m (30-33 ft.-lbs.) torque.

Models With Tandem Pump

29. REMOVE AND REINSTALL. Clean pump and area around pump thoroughly, then disconnect lines from pump and allow fluid to drain. Cover all openings to prevent dirt from entering pump or lines, then unbolt and remove pump from the engine timing case.

Tighten pump retaining screws to 31-42 N·m (23-31 ft.-lbs.) torque. Refer to paragraph 22 for filling and bleeding.

30. OVERHAUL. Several different pumps have been used on these models. Refer to the following and Fig. 42, Fig. 43 and Fig. 44 for typical service.

Straighten tab of lock washer, remove nut, then use a puller to remove the pump drive gear. Remove Woodruff key after gear is pulled from shaft. Mark relative position of flange housing (11), bodies (2 and 10) and covers (1, 3 and 9), then remove bolts (12). Pump gears (4 and 16) are available only as matched set. Check condition of bearing blocks (5 and 17), gears (4 and 16) and bodies (2 and 10) for wear or other damage.

On Bosch pump (Fig. 42), mark location of each part of bearings (5 and 17) to be sure parts are reinstalled in same location. Grooves less than 0.3 mm (0.012 inch) deep are permissible in the larger (inlet) hole of pump bodies (2 and 10). Check and install new inner seals (6 and 18) as follows. Height of seal support strip should be 1.95-2.10 mm (0.077-0.085 inch). If any part of support strip is too high, turn strip upside down in a cover and carefully polish

Fig. 41—Exploded view of separate steering pressure pump used on some three-cylinder models. Unit shown is made by Bendix (Hydraquip). Refer also to Fig. 40 for other separate pump.

1. Nut
2. Reservoir
3. Filter element
4. "O" ring
5. Spring
6. Fill plug
7. Gasket
8. Gasket
9. Cover
10. Pump body
11. Cover and flange
12. Screw
14. Seal
16. Pump gears
17. Bearings
18. Inner seals
19. Outer seal
20. Seal rings
21. Washer

Fig. 42—Exploded view of Bosch tandem pump used on some models.

1. Cover
2. Pump body
3. Cover
4. Pump gears
5. Bearings
6. Inner seals
7. Outer seal
8. Coupling
9. Cover
10. Pump body
11. Cover and flange
12. Screw
13. Snap ring
14. Seal
15. Woodruff key
16. Pump gears
17. Bearings
18. Inner seals
19. Outer seal

Fig. 43—Exploded view of Sundstrand tandem pump used on some models.

1. Pump center housing
2. Pump body
4. Pump gears
5. Bearings
6. Inner seals
8. Coupling
9. Cover
10. Pump body
11. Cover and flange
12. Screw
14. Seal
16. Pump gears
17. Bearings
19. Outer seal

Fig. 44—Exploded view of Cessna tandem pump used on some models.

1. Pump center housing
2. Pump body
4. Pump gears
9. Cover
10. Pump body
11. Cover and flange
12. Screw
14. Seal
15. Woodruff key
16. Pump gears
18. Inner seals
19. Outer seals
21. Washer
22. Wear plate

lightly with emery paper. Install support strips and seal strips in covers with beveled edge of support strips and rounded edge of seal strips against rounded (radius) sides of seal groove. Marks affixed to parts before disassembling should be aligned when reassembling. Support strip is toward suction side and seal strip is toward pressure side. Tighten M8 screws (12) to 17-23 N·m (13-17 ft.-lbs.) torque and M10 screws to 35-46 N·m (26-34 ft.-lbs.) torque. Tighten screws evenly and check for free rotation of pump shaft while tightening. Tighten pump drive gear retaining nut to 40-45 N·m (30-33 ft.-lbs.) torque.

On Sundstrand pumps (Fig. 43), bearing (5) nearest the cover with mounting flange (11) is thicker than other bearings and extends into front cover. Pack shaft seal (14) with grease before installing and stick seals (6, 7, 18 and 19) into position with grease while assembling. Marks affixed to parts before disassembling should be aligned when reassembling. Tghten screws (12) on pumps used on six-cylinder engines to 61-68 N·m (45-50 ft.-lbs.) torque. Tighten screws evenly and check for free rotation of pump shaft while tightening. Tighten pump drive gear retaining nut to 40-45 N·m (30-33 ft.-lbs.) torque.

On Cessna pumps (Fig. 44), auxiliary pump gears (4) should be at least 12.95 mm (0.510 inch) thick and steering pump gears (26) should be at least 9.75 mm (0.384 inch) thick. Shaft bushings are not available separately from covers (1, 9 and 11). If inside diameter of bushings exceeds 19.177 mm (0.755 inch), install new covers and bushing assemblies. Inside diameter of gear pockets in bodies (2 and 10) should be 43.51 mm (1.713 inch) or less. The two half moon cavities in pump bodies (2 and 10) must face away from front plate (11) and the smaller half moon cavity must be on pressure side of pump. Install wear plate (22) with bronze side toward gears and larger cut-away in middle section toward suction side of pump. Marks affixed to parts before disassembling should be aligned when reassembling. Install washer (21) and seal (14) after pump is assembled and screws (12) are installed and tightened to 37-40 N·m (27-30 ft.-lbs.) torque. Tighten screws evenly and check for free rotation of pump shaft while tightening. Lubricate seal (14), washer (21) and pump drive shaft, install washer (21), position a 33 mm (1.3125 inches) diameter seal protector over shaft, then install seal (14) being careful not to damage seal lip. Press seal into front cover until seal is 0.5-1.0 mm (0.020-0.040 inch) below front face of cover (11). Tighten pump drive gear retaining nut to 40-45 N·m (30-33 ft.-lbs.) torque.

To run in a new or rebuilt pump, install pump and run engine for three minutes at 1200 rpm at 0 (zero) hydraulic pressure, only circulating hydraulic fluid. Operate auxiliary control valve intermittently (to build pressure) for three minutes, then cycle the

steering wheel lock to lock for three minutes. Increase engine speed to rated rpm and recheck auxiliary and steering systems. Idle engine and check for leaks. Change filter and refill transmission oil.

POWER CYLINDER

Two-Wheel-Drive Models With Three-Cylinder Engine

31. R&R AND OVERHAUL. One cylinder is attached between axle center member and each steering arm as shown in Fig. 45. To remove the cylinder, disconnect hydraulic hoses at steering cylinder and cover openings to prevent entrance of dirt, then detach both ends of the steering cylinder.

Clamp steering cylinder in a vise or holding fixture, unscrew end cap (1—Fig. 46), then withdraw piston (3) and rod (2) from the cylinder.

When assembling, apply "Loctite 270" on threads of piston (3) and rod (2), tighten piston, then stake threads with center punch to prevent loosening.

Tighten nut securing rod end to steering arm to 75-90 N·m (55-66 ft.-lbs.) torque. Remainder of installation follows standard procedures. Cycle the system several times and check for leakage.

Two-Wheel-Drive Models With Six-Cylinder Engine

32. R&R AND OVERHAUL. The cylinder is attached between axle and steering arms as shown in Fig. 47. To remove the cylinder, first refer to paragraph 4 and remove the axle and steering cylinder as an assembly. The steering cylinder is attached to the

Fig. 45—Installed view of one steering cylinder typical of all tractors with a three-cylinder engine and two-wheel drive.

Fig. 46—Exploded view of steering cylinder used on two-wheel-drive tractors with three-cylinder engines.

1. End cap
2. Piston rod and rod end
3. Piston
4. "O" ring & sealing ring
5. Cylinder end housing
6. Pressure seal
7. "O" ring
8. Washer
9. Dust seal
10. Cylinder
11. Sealing ring
12. Steering cylinder housing

axle center member with pivot block (8—Fig. 8). Remove cotter pin, remove nut, then bump tapered end of ball-joint (3) from pivot block (8). Complete removal of steering cylinder by detaching rod ends from steering arms (1).

Refer to Fig. 8 and remove the track rods (6) and rod ends. Ball-joint (3) can be removed after removing the clamp screw. Clamp the cylinder (1—Fig. 48) in a holding fixture and clean sealing compound from elongated hole (H). End cap (7) has two holes which can be used to rotate the end cap until the end of the retaining ring (2) is located in the hole. Use a small screwdriver or similar instrument to pull and guide the end of the retaining ring out through the elongated hole (H). Turn the end cap (7) and pull the retaining ring from between the end cap and cylinder. One end of the retaining ring is bent so that end enters a hole in the end cap. After retaining ring (2) is removed, withdraw piston and rod (12), cap (7) and seals from cylinder (1). Remove old seals and install new seals before reassembling.

Lubricate and assemble steering cylinder by reversing disassembly procedure. Turn end cap (7) in cylinder until the hole in end cap for end of the

retaining ring (2) is visible in hole (H). Insert the bent end of retaining ring in hole and turn end cap, drawing retaining ring (2) into groove formed between end

Fig. 48—Exploded view of the steering cylinder used on two-wheel-drive, six-cylinder tractors. Sealing compound is used to seal hole (H) after assembly and installation of retaining ring (2).

1. Steering cylinder	7. End cap
2. Retaining ring	8. Wear ring
3. Seal ring	9. Piston wear ring
4. "O" ring	10. Seal ring
5. Dust seal	11. "O" ring
6. "O" ring	12. Piston and rod assembly

Fig. 47—Steering cylinder can be removed from two-wheel-drive, six-cylinder tractors after first removing axle assembly.

cap (7) and cylinder (1). Cover hole (H) with sealing compound after the retaining ring is installed to prevent the entrance of water that would rust the end cap and the retaining ring. If removed, install ball-joint (3—Fig. 8) after applying "Loctite 270" to threads. Tighten ball-joint to 80-90 N•m (59-66 ft.-lbs.) torque and ball-joint clamp bolt to 120-160 N•m (88-96 ft.-lbs.) torque. Tighten slotted nut retaining ball-joint in block (8) to 80-140 N•m (60-103 ft.-lbs.) torque, then install cotter pin. Side play of block (8) between bracket arms of axle should be 0-0.2 mm (0-0.008 inch) and is controlled by the thickness of shims (23). Tighten bolts attaching track rods (6) to piston rod to 29-37 N•m (21-27 ft.-lbs.) torque. Tighten slotted nuts attaching rod ends (2) to the steering arms (1) to 50-70 N•m (37-52 ft.-lbs.) torque, then install cotter pins. Adjust position of steering hose elbows for clearance between front support, without hanging down below axle. Make sure hoses do not restrict tipping movement of axle after installation. Refer to paragraph 4 when reinstalling axle.

All Four-Wheel-Drive Models

33. R&R AND OVERHAUL. The cylinder is attached between the axle center housing and the steering arm as shown in Fig. 49. To remove the cylinder, disconnect hydraulic hoses at steering cylinder and cover openings to prevent entrance of dirt, then detach both ends of the steering cylinder.

Clamp the cylinder (1—Fig. 50) in a holding fixture and clean sealing compound from elongated hole (H). Rotate the end cap (4) until the end of the retaining ring (2) is located in the hole. Use a small screwdriver or similar instrument to pull and guide the end of the retaining ring out through the elongated hole (H). Turn the end cap (4) and pull the retaining ring from between the end cap and cylinder. One end of the

retaining ring is bent so that end enters a hole in the end cap. After retaining ring (2) is removed, withdraw piston rod (13), piston (11), cap (4) and seals from cylinder (1). Rod end and/or nut (7) can be removed, but if rod end is removed, measure position of old rod end so that new rod end can be installed at same location on rod (13). Remove all old seals and install new seals before reassembling.

Lubricate and assemble steering cylinder by reversing disassembly procedure. Turn end cap (4) in cylinder until the hole in end cap for end of the retaining ring (2) is visible in hole (H). Insert the bent end of retaining ring in hole and turn end cap, drawing retaining ring (2) into groove formed between end cap (4) and cylinder (1). Cover hole (H) with sealing compound after the retaining ring is installed to prevent the entrance of water that would rust the end cap and the retaining ring.

Measure distance between the center of the pivot pin to the center of the ball end when cylinder is completely retracted. Correct distance should be as follows.

Models	Length
340, 355	418 mm (16.469 inch)
360	418 mm (16.469 inch)
399	462 mm (18.203 inch)

Tighten ball-joint to 98-108 N•m (72-80 ft.-lbs.) torque for 340, 355 and 360 models; tighten to 177-196 N•m (130-145 ft.-lbs.) torque for 399 models. Adjust position of steering hose elbows for clearance between front support, without hanging down below axle. Make sure hoses do not restrict tipping movement of axle after installation.

Fig. 50—Exploded view of steering cylinder used on four-wheel-drive tractors.

1. Cylinder
2. Retaining ring
3. "O" ring
4. End cap
5. Seal
6. Seal
7. Nut
8. Wear ring
9. Seal ring
10. "O" ring
11. Piston
12. "O" ring
13. Piston rod

Fig. 49—Installed view of steering cylinder typical of four-wheel-drive tractors.

STEERING CONTROL VALVE

All Models

34. R&R AND OVERHAUL. The steering control valve is located at the base of the steering wheel shaft

Fig. 51—View of four screws attaching steering control valve. Some models have standard head screws, while other models are fitted with socket head screws as shown.

and removal from tractors with cab requires tractor to be split between the engine and the gearbox as outlined in paragraph 152. Removal is possible without splitting on tractors without cab, but hoses must be disconnected and clearance is very tight. Remove cover from base of steering column, remove the four screws (S—Fig. 51), lower the steering control valve off steering shaft splines, then remove unit from tractor.

On all models, thoroughly clean exterior of unit. Remove the seven cover retaining screws (10—Fig. 52), then remove the cover (9), stator (7), rotor (6), spacer ring (2) and "O" rings (5 and 8). Remove distributor plate (4), drive shaft (1) and "O" ring (3). Hold steering valve vertically and turn valve spool and sleeve to align cross pin (20) parallel to flat side of housing. With cross pin in this position and housing in horizontal position, remove sleeve (22), spool (21), thrust bearing (18) and bearing races (17 and 19) from housing. Remove cross pin (20) from rotary valve and separate spool (21) from sleeve (22). Remove leaf springs (23) from spool.

Do not attempt to remove shock valves from steering cylinder ports (F and G—Fig. 58). The non-return valve in the inlet port (D) should not require servicing either. If necessary, the pressure relief valve (32 through 36—Fig. 52) can be removed after unscrewing plugs (32 and 34); however, pres-

Fig. 52—Exploded view of steering control valve with integral system relief valve (34-36).

1. Drive shaft	13. Check valve ball	26. Valve seat
2. Spacer	14. Check valve plug	27. "O" ring
3. "O" ring	15. Housing	28. Check ball
4. Distributor plate	17. Bearing race	29. Ball stop
5. "O" ring	18. Thrust bearing	30. Back-up ring
6. Rotor	19. Bearing race	31. "O" ring
7. Stator	20. Cross pin	32. Plug
8. "O" ring	21. Valve spool	33. "O" ring
9. End cover	22. Valve sleeve	34. Adjusting plug
10. Cap screw (7)	23. Leaf springs (6)	35. Relief valve
11. Oil seal	24. Retainer	spring
13. Check valve ball	25. "O" ring	36. Relief valve

Fig. 53—Marks on spool and sleeve indicated by arrow and slot for centering springs should be aligned as shown when assembling.

Fig. 54—Arch of the four center leaf springs (23) should be together and the flat springs (23S) should be on the outside as shown.

Fig. 55—Check valve ball (13—Fig. 52) should be in location indicated.

sure must be checked and adjusted if plug (34) is turned. Steering relief pressure is checked as outlined in paragraph 24 or paragraph 25 and adjusted by turning plug (34).

Clean and inspect all parts for excessive wear or other damage and renew parts as necessary. Housing (15), spool (21) and sleeve (22) are not available separately. Use all new "O" rings and seals when reassembling. Lubricate all interior parts with clean steering fluid.

Insert spool (21) into sleeve (22) aligning the two marks (Arrow—Fig. 53) and leaf spring slots. Install leaf springs (23), with two flat springs on the outside and two sets of two arched springs in the middle as shown in Fig. 54. Special tool MS63 can be used when installing leaf springs. Insert cross pin (20—Fig. 52)

into sleeve and spool and install retainer (24) over the leaf springs. Press seal (11) in housing (15) until flush. Install thrust bearing (17, 18 and 19) on spool, making sure that chamfered side of race (17) is away from needle bearing (18). Use special tool MS62A to install the "O" ring and back-up ring into position in housing (15), then carefully insert the assembled and lubricated spool in housing. The cross pin should be kept parallel with the flat surface of housing near the hose connections.

Insert the check valve ball (13) and plug (14) into the hole indicated in Fig. 55. Lubricate the "O" ring (3—Fig. 52) and install in groove of housing, followed by valve plate (4). Make sure that all of the holes in valve plate and housing are aligned. Turn the valve spool assembly so that pin (20) is parallel with the flat port face of housing (15) where hoses attach. Mark a reference line across splined end of rotor shaft (1) parallel with the groove at other end for pin. Be sure that valve spool pin is still aligned with the port face of the housing and install the rotor shaft as shown in Fig. 56. When correctly assembled, one lobe of the rotor will be positioned straight away from the port face (up in Fig. 56) when pin center line is located as shown. Install "O" ring (5—Fig. 52), pump body (7) and pump rotor (6). Install spacer (2), "O" ring (8) and end cover (9). Install the seven retaining screws (10). The one screw with the pin attached should be in location identified as "7" in Fig. 57. Tighten the screws in the order shown in Fig. 57, first to 10-15 N·m (7-10 ft.-lbs.), then to 25-35 N·m (19-26 ft.-lbs.) torque. Install relief valve assembly (32 through 36— Fig. 52) if removed. If setting of plug (34) has been disturbed, refer to paragraph 24 or paragraph 25 for

Fig. 56—The center line of pin (20—Fig. 52) should be aligned parallel with the port face of housing as shown when assembling. Center line of pin is below center line of shaft.

Fig. 57—The one screw with pin attached should be installed in position (7). Numbers indicate the order which screws should be tightened.

Fig. 58—Groove in splined end of coupling (B) should be level with top of steering shaft. Hose from steering pump should be attached to right angle fitting (D). Hose back to pump or to the oil cooler should be attached to fitting (E). Hose to left steering cylinder should be attached at fitting (F) and hose to right steering cylinder should be attached to fitting (G).

checking and adjusting relief valve pressure. Tighten plugs (24 and 32) to 50 N·m (37 ft.-lbs.) torque.

Installation is reverse of removal procedure. Groove (B—Fig. 58) in splined end of flexible coupling should be level with top of steering unit drive shaft. If screws (M—Fig. 51) or steering wheel retaining nut (C—Fig. 58) were removed, tighten to 34-37 N·m (25-35 ft.-lbs.) torque. Refer to Fig. 58 for proper attachment of steering hoses (D, E, F and G).

THREE-CYLINDER ENGINE AND COMPONENTS

Three-cylinder engines are installed in 340, 350, 355 and 360 models. Refer to paragraph 67 and following for service to six-cylinder (6.3544) engines used in early 399 models. Refer to paragraph 96 and following for service to six-cylinder (1006.6) engines used in late 399

models. Refer to Condensed Service Data for original application. The engine serial number consists of up to 13 letters and numbers located as shown in Fig. 59. The numbers and letters constituting the engine serial number can be used to identify the engine family, country of manufacture and the year of manufacture. Refer to Fig. 60 for identifying code.

Fig. 59—View of three engine cylinder showing location of the engine serial number.

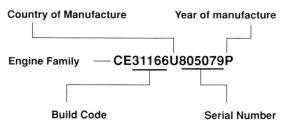

Fig. 60—The first two letters of the engine serial number identifies engine family, the second group is the build code. The letter approximately in the center identifies the country of manufacture and is followed by the engine serial number. The last letter indicates the year of manufacture. The illustration shown above identifies a D3.152 engine (naturally aspirated, three-cylinder engine with 152 cubic inch displacement), built in the United Kingdom in 1987. The build code is 31166 and the serial number is 805079.

Engine family –
CE = D3.152 (Naturally aspirated, 3 cyl., 152 cid)
CN = T3.1524 (Turbocharged, 3 cyl., 152 cid)
Country of manufacture –
A = Argentina
B = Brazil
C = Australia
D = Germany
E = Spain
F = France
G = Greece
J = Japan
K = Korea
L = Italy
M = Mexico

N = U.S.A.
P = Poland
S = India
T = Turkey
U = United Kingdom
W = Iran
X = Peru
Y = Yugoslavia
Year of manufacture –
N = 1986
P = 1987
S = 1988
T = 1989
U = 1990
V = 1991
W = 1992

FRONT END/ENGINE SPLIT

Three-Cylinder Models

35. Block both rear wheels to prevent rolling, remove any front mounted equipment and any front weights. Remove hood, grille and side panels and drain coolant from radiator and engine block. Disconnect both battery cables at the battery, then free the positive cable so that it can stay with the engine. Disconnect front wiring harness and thread it back so that it can stay with the engine. Drain fuel tank, if necessary, then disconnect fuel feed and return lines. Remove both tank side supports, then lift fuel tank from tractor. Disconnect top and bottom radiator hoses, hydraulic pipes to and from the oil cooler, hoses to the steering cylinders and the hose between the air filter and the inlet manifold. Disconnect pipe to turbocharger, if fitted. On four-wheel-drive tractors, refer to paragraph 9 and remove drive shaft and shield. On all models, wedge between the front support and axle to prevent tipping and support the front axle, radiator and nose from above using a movable hoist. Support rear of tractor under transmission housing, then carefully remove stud nuts and screws attaching front support to the engine. Carefully roll the front axle, radiator and front assembly forward, away from engine as shown in Fig. 61.

Reassemble by reversing the splitting procedure. Refer to paragraph 5 for attaching front support to engine and selecting proper shims. Refer to paragraph 9 for installing drive shaft of four-wheel-drive models.

R & R ENGINE

Three-Cylinder Models

36. To remove the engine, first refer to paragraph 35 and split tractor between front support and front of engine. Disconnect wires from engine components and position wires out of the way. Disconnect throttle and stop cables from the fuel injection pump, detach cables from support brackets and move the cables out of the way. Disconnect hydraulic lines between rear

Fig. 61—View of two-wheel-drive front axle, radiator and associated parts separated from the front of three-cylinder engine. Use care when supporting both front and rear sections to prevent tipping or other unwanted movement.

of tractor and front that would interfere with the removal of engine and cover all openings. On models with cab, disconnect heater and air conditioning lines if so equipped. Remove gear lever trim plate and boot from models with cab. On all models, attach hoist to the engine, then remove the screws and nuts attaching the engine to transmission housing. Carefully move the engine forward away from the transmission.

To reinstall engine, reverse the removal procedure and observe the following. Install a M12 alignment stud approximately 100 mm (4 inches) long in each side of the transmission housing to assist in aligning engine to transmission. Turn flywheel to align clutch plate splines with transmission and pto input shaft splines. Install retaining screws and nuts after engine and transmission housing flanges are completely together. Tighten retaining screws and nuts to 115 N·m (85 ft.-lbs.) torque. Reverse removal procedure and refer to paragraph 5 for attaching front support to engine and selecting proper shims. Check and adjust clutch linkage as outlined in paragraph 151. Refer to paragraph 9 for installing drive shaft of four-wheel-drive models.

CYLINDER HEAD

Three-Cylinder Models

37. REMOVE AND REINSTALL. To remove cylinder head, first drain cooling system, remove hood and disconnect battery ground cable. Disconnect exhaust pipe from manifold or turbocharger and inlet pipe from manifold or turbocharger. Remove turbocharger and crossover pipe from models so equipped. Remove the fuel pressure and return lines from all

models. Disconnect breather pipe and top radiator hose, then remove rocker arm cover. Detach exterior oil line and rocker arm lubricating oil line from cylinder head, then unbolt and remove the rocker arm assembly. Unbolt and remove inlet and exhaust manifolds from cylinder head. Unbolt and remove fuel injector assemblies from cylinder head. On some models, the cylinder head can be more easily removed if the coolant outlet housing is first removed. Loosen screws and stud nuts retaining cylinder head in reverse of order shown in Fig. 62, then remove fasteners and lift off cylinder head. **Do not use sharp or hard tools between cylinder block and head, and lay removed head on clean flat surface to prevent damage.**

Refer to specific paragraphs for servicing rocker arms, valves, seats and guides. Clean gasket surface of cylinder head and check for distortion or other damage. Check flatness of cylinder head gasket surface using a straightedge and feeler gauge. Cylinder head should be renewed or resurfaced if surface is

Fig. 62—On three-cylinder models, tighten cylinder head retaining screws in the order shown. Removal should be in reverse order shown to prevent damage.

warped more than 0.076 mm (0.003 inch) across head or 0.152 mm (0.006 inch) from front to rear. No more than 0.30 mm (0.012 inch) can be removed from head, and nozzle protrusion should be less than 4.55 mm (0.179 inch).

Different gaskets and fasteners have been used and are available for service. Cylinder head retaining screws and nuts should be the later type which can be recognized by the integral washers. The cylinder head retaining screws and stud nuts should be tightened to 95 N·m (70 ft.-lbs.) torque. Cylinder head retaining fasteners do not require retorquing if the latest type cylinder head gasket and the late type fasteners (with integral washer) are used. Nuts retaining injection nozzles should be tightened to 16 N·m (12 ft.-lbs.) torque. Coolant outlet (thermostat) housing should be tightened to 13.6 N·m (10 ft.-lbs.) torque.

Refer to paragraph 42 when installing rocker arm assembly. Check and adjust valve clearance as outlined in paragraph 43. Remainder of reassembly procedure will be reverse of removal procedure.

VALVES AND SEATS

Three-Cylinder Models

38. When disassembling the cylinder head, keep valves in order so that they can be reinstalled in original location, if reused. Valve heads are marked with cylinder numbers consecutively from front to rear on some engines. Any replacement valves should be marked before installation.

Inlet and exhaust valves originally seat directly in cylinder head of most models; however, cylinder head may be machined and fitted with replacement seats. Inlet valve face angle is 35° and inlet seat angle is 36° for all models. Exhaust valve face angle is also 35° and seat angle is 36° for all models except 360 model. Inlet valves for Model 360 also have 35° face angle and 36° seat angle, but the exhaust valves have face angle of 45° and seats are 46°. The inlet valves of all models should be recessed 1.32-1.63 mm (0.052-0.064 inch) below surface of cylinder head and should not exceed 2.26 mm (0.089 inch). The exhaust valves of all models should be recessed 1.60-1.92 mm (0.063-0.075 inch) below surface of cylinder head and should not exceed 2.16 mm (0.085 inch). Valve stem diameter is 7.90-7.92 mm (0.3110-0.3118 inch) for all valves and all models.

VALVE GUIDES

Three-Cylinder Models

39. Standard inside diameter of inlet and exhaust valve guides is 7.98-8.01 mm (0.3141-0.3155 inch). New valve stem diameter is 7.90-7.92 mm (0.311-

0.312 inch) for all valves and desired stem to guide clearance is 0.05-0.11 mm (0.0021-0.0045 inch). Renew valve and/or guide if clearance exceeds 0.15 mm (0.006 inch) for inlet valve or 0.14 mm (0.0055 inch) for exhaust valve. To remove worn guides, press guides from top using suitable removing tools. The guides are interchangeable on some models, but the inlet guide is 5.5 mm (7/32 inch) shorter than exhaust on turbocharged engines. The longer guide should be 14.73-15.09 mm (0.580-0.594 inch) above the spring seat when properly installed. The shorter guide, used for the inlet valve of some models, should extend 9.19-9.55 mm (0.362-0.376 inch) above the spring seat. Guides are presized and should not require resizing if carefully installed using the proper tools. Reface valve seats after renewing guides to obtain concentricity of seat to guide.

VALVE SPRINGS

Three-Cylinder Models

40. Some models are fitted with inner and outer valve springs on both inlet and exhaust valves, but other models use inner springs only on inlet valves. Close wound damper coils should be installed toward cylinder head. Renew springs if discolored, distorted or if they fail to meet the following specifications. Outer spring free length is 45.29-45.80 mm (1.783-1.803 inches) and spring should exert 9.44-11.24 kg (20.81-24.78 lbs.) when compressed to 30.16 mm (1.1875 inches). Inner spring free length is 34.67-35.69 mm (1.365-1.405 inches) and spring should exert 3.18-4.08 kg (7.01-8.99 lbs.) when compressed to 30.16 mm (1.1875 inches).

CAM FOLLOWERS

Three-Cylinder Models

41. The mushroom type cam followers operate directly in machined bores in cylinder head. Refer to Fig. 63. With cylinder head removed, cam followers can be removed after first removing the adjusting screw and locknut from top of cam follower. Identify cam followers as they are removed so each can be reinstalled in original location if reused. Cam follower stem diameter is 15.80-15.84 mm (0.6222-0.6237 inch) and stems should have 0.02-0.09 mm (0.0008-0.0036 inch) clearance in cylinder head bores.

ROCKER ARMS

Three-Cylinder Models

42. The rocker arms and shaft assembly can be removed after removing the hood, fuel tank and

Fig. 63—Mushroom cam followers protrude from under side of cylinder head as shown.

Fig. 64—Early rocker shaft assemblies should be installed with slot (S) in end of rocker arm shaft toward front as shown. Some models are located by pin shown in Fig. 65.

rocker arm cover. The rocker arms are right and left hand units and should be assembled as shown in Fig. 66.

Rocker shaft diameter is 15.81-15.84 mm (0.6223-0.6238 inch) and the desired diametral clearance between shaft and rocker arms is 0.03-0.10 mm (0.0010-0.0038 inch). Renew shaft and/or rocker arms if clearance exceeds 0.13 mm (0.005 inch) or if either surface of shaft or bushing in rocker arm is rough. The amount of oil circulated to the rocker arms is regulated by the position of the rocker arm shaft in the brackets. Correct shaft position for early models is with slot (S—Fig. 64) in end of shaft located at

Fig. 65—Late rocker shaft assemblies are located by tapered pin (P) in rocker arm shaft that should fit into slot at rear of second bracket as shown.

about 30° from vertical as shown. Correct shaft position for later models is with pin (P—Fig. 65) in slot of second shaft bracket. Studs retaining rocker shaft brackets are shorter and holes in cylinder head are not tapped as deep in later models. Be sure that correct studs are provided with head used.

VALVE CLEARANCE

Three-Cylinder Models

43. Recommended valve clearance on 340 and 350 models is 0.3 mm (0.012 inch) cold for both inlet and exhaust valves. Recommended clearance for turbocharged 355 and 360 models is 0.3 mm (0.012 inch) cold for exhaust valves and 0.2 mm (0.008 inch) for inlet valves. Static setting of all valves can be made from two crankshaft positions using the following procedure:

Remove the timing window plug from the left side of flywheel adapter housing and turn the flywheel until "TDC 1" mark on flywheel is aligned in center of hole. Check valves for front cylinder. Both valves should be closed if front cylinder is on compression stroke. If front two valves are tight (open), turn crankshaft exactly one complete revolution to set front piston on compression stroke.

If not equipped with hole in flywheel adapter or mark on flywheel, turn flywheel in normal direction of rotation until the exhaust valve for the front cylinder is just about closed and the inlet valve is just beginning to open. If not equipped with marks, mark crankshaft pulley so that crankshaft can be rotated exactly one complete revolution from this position after adjusting the first set of valves indicated in Fig. 67.

Fig. 66—With TDC timing marks aligned and number 1 (front) piston on compression stroke, adjust the valves indicated. After adjusting these valves, turn crankshaft exactly one revolution (until marks again align) then adjust valves shown in Fig. 67.

Fig. 67—With TDC timing marks aligned and number 1 (front) piston on exhaust stroke, adjust the valves indicated.

With front cylinder set at TDC on compression stroke, adjust the four valves indicated in Fig. 66. Turn the crankshaft exactly one complete revolution and adjust the two valves indicated in Fig. 67.

VALVE TIMING

Three-Cylinder Models

44. Timing gears are correctly timed if marks are properly aligned as outlined in paragraph 51. Valve timing is not adjustable and gears can only be improperly timed by improper assembly or damage to the gear teeth. Refer to paragraph 45 for removing the timing gear cover, then to paragraph 51 for checking and setting the timing marks.

TIMING GEAR COVER

Three-Cylinder Models

45. To remove the timing gear cover, refer to paragraph 35 and remove the front axle, support and associated parts. Remove accessory drive belts, alternator, coolant pump and crankshaft pulley. Remove screws attaching the timing gear cover to the front of engine. Note that two of the long screws which attach timing cover also attach alternator bracket. The lower screw, below the crankshaft oil seal, is fitted with an aluminum or copper sealing washer. Be careful to be sure that all of the screws are removed. The camshaft thrust spring is riveted to timing gear cover and controls end play of the camshaft.

Press (do not pry) old crankshaft oil seal from cover bore, then press new seal into position with spring loaded lip toward inside. Press new seal into bore until front surface of seal case is 11.43 mm (0.450 inch) from front surface of timing gear cover of 340 and 350 models; or 15.88 mm (0.625 inch) for 355 and 360 models.

To install timing gear cover, first install a new seal into cover, then position gasket on front of engine. Position cover and install retaining screws loosely. Center timing gear cover around the crankshaft using special centering tool (PD159 available from Massey-Ferguson) or equivalent, then tighten the retaining screws. If special centralizing tool is not available, the crankshaft pulley can be used carefully, but damage to the soft lip of the seal is likely. Tighten crankshaft pulley retaining screw to 142 N·m (105 ft.-lbs.) torque if washer is 4.8 mm (0.19 inch) thick; to 325 N·m (240 ft.-lbs.) torque if washer is 8.9 mm (0.35 inch) thick.

TIMING GEARS

Three-Cylinder Models

46. Before attempting to remove any of the timing gears, first remove the fuel tank, rocker arm cover and the rocker arms to avoid possible damage to pistons or valve train. The valves can (and probably will) hit the pistons if either the camshaft or crankshaft is turned independently of the other. The timing marks shown in Fig. 68 align only once every 18 revolutions due to the odd number of teeth on the idler gear.

To remove the timing gears or align marks on gears, refer to paragraph 35 and remove the front axle, support and associated parts, then remove the timing gear cover as outlined in paragraph 45. Unstake and remove the idler gear retaining screw and slide gear from idler shaft. The shaft is a light press fit in timing gear housing and is further located by the pin shown

Fig. 68—View of timing gears with marks aligned. Marks on the crankshaft, camshaft and injection pump gears will align with respective marks on idler gear as shown only once every 18 revolutions.

Fig. 69—The oil holes in cylinder block and idler gear stub shaft are aligned by the locating dowel pin.

in Fig. 69. Remove idler shaft by prying gently from bore and locating pin. Refer to paragraphs 47, 48, 49 and 50 for service information regarding each of the four gears. With rocker arms removed, crankshaft gear, camshaft gear and injection pump gear can be rotated as necessary before reinstalling idler gear with all timing marks aligned.

47. IDLER GEAR. The idler gear is retained on idler shaft by a cap screw and washer. Idler gear bore diameter is 53.98-54.02 mm (2.1250-2.1266 inches) and idler shaft diameter should be 53.92-53.94 mm (2.1230-2.1238 inches). Gear to shaft diametral clear-

ance should be 0.03-0.09 mm (0.0012-0.0036 inch). Backlash should be at least 0.08 mm (0.003 inch) between the large idler and any other gear. Gear should have 0.13-0.38 mm (0.005-0.015 inch) end play on shaft.

Make sure that timing marks on all gears are aligned as shown in Fig. 68 when idler gear is installed. Tighten retaining screw to 68 N·m (50 ft.-lbs.) torque, then bend locking washer against a flat of the screw to prevent loosening.

48. CAMSHAFT GEAR. The camshaft gear is attached to the camshaft by three equally spaced cap screws. It is therefore possible to attach the camshaft gear to the camshaft three different ways, but only one is correct. To correctly install the gear, align the stamped letters "D" on hub of camshaft and front face of gear as shown at (X—Fig. 68). Tighten gear retaining cap screws to 28 N·m (21 ft.-lbs.) torque.

49. CRANKSHAFT GEAR. The crankshaft gear is keyed in place and fits the shaft with a transitional fit, 0.03 mm (0.001 inch) tight to 0.03 mm (0.001 inch) loose. If the old gear is loose on shaft, it may be possible to pry gear from shaft, but a puller is usually required. If necessary to remove crankshaft gear with a puller, the oil pan and lower section of the timing gear housing must first be removed.

When installing gear, be sure that marked tooth is toward front. Install washers and oil slingers as originally fitted behind and in front of crankshaft gear. Use a straightedge to align front face of timing gear housing lower section with front face of housing.

50. INJECTION PUMP DRIVE GEAR. The injection pump drive gear is retained to the pump adapter by either three cap screws or a center nut. When installing the gear, align dowel pin (1—Fig. 70) with slot (2) in adapter hub or drive key with slot, then install the retaining screws or nut. The injection

Fig. 70—Dowel pin (1) located in the injection pump drive gear must enter machined slot (2) in adapter shaft.

pump drive gear and adapter are supported by the injection pump rotor bearings.

51. TIMING THE GEARS. The crankshaft gear has 25 teeth, the camshaft gear has 50 teeth and the injection pump drive gear has 50 teeth, but because the idler gear has 90 teeth, marks on the four gears will align only every 18 crankshaft revolutions. To align the marks, it is easier and faster to turn the crankshaft until the marked teeth of the crankshaft, camshaft and injection pump gears are in a position as shown in Fig. 68. Remove the idler gear, then reinstall with all of the marks aligned.

Do not turn the crankshaft or camshaft with the idler gear removed, because the valves can (and probably will) hit the pistons if either the camshaft or crankshaft is turned independently of the other. If it is necessary to turn either the camshaft or crankshaft to align timing marks, first remove the fuel tank, rocker arm cover and the rocker arms to avoid possible damage to pistons or valve train.

If rocker arms have been removed, turn the crankshaft, camshaft and injection pump gears to approximately the positions shown in Fig. 68, then install the idler gear with all of the marks aligned.

TIMING GEAR HOUSING

Three-Cylinder Models

52. To remove the timing gear housing, refer to paragraph 35 and remove the front axle, support and associated parts, then refer to paragraph 45 and remove the timing gear cover. Remove the fuel tank, rocker arm cover and rocker arm assembly. Refer to paragraph 46 and remove the idler gear, remove camshaft as outlined in paragraph 53 and injection pump drive gear as outlined in paragraph 50. Remove fuel lift pump, the oil pan and the lower section of the timing gear housing. Remove the steering pump and the fuel injection pump from rear surface of timing gear housing. Unbolt and remove the timing gear housing from front face of cylinder block.

Reinstall timing gear housing by reversing the removal procedure. Use a straightedge to align front face of timing gear housing lower section with front face of housing. Refer to paragraph 51 and Fig. 68 for aligning the timing marks.

CAMSHAFT

Three-Cylinder Models

53. To remove the camshaft, refer to paragraph 35 and remove the front axle, support and associated parts, then refer to paragraph 45 and remove the timing gear cover. Remove the fuel tank, rocker arm cover and rocker arm assembly. Raise cam followers away from camshaft and secure in the fully raised position, then withdraw the camshaft as shown in Fig. 71.

The camshaft operates in three journal bores machined in the cylinder block. The front camshaft journal bore may be fitted with a bushing. The center journal of the camshaft is pressure lubricated by oil that passes through an external oil line. The camshaft center journal, in turn, meters oil to the rocker arm shaft and cylinder head through a second oil line between the cylinder head and the rocker shaft. Some oil is metered from the center journal to the front journal bearing. The rear camshaft bearings are lubricated by oil returning by gravity after first lubricating the rocker arms and valves. The lobes of the camshaft are lubricated by oil splashed and drained from other locations.

Inspect camshaft lobes and bearing journals for excessive wear, scoring, pitting or other damage. Refer to the following specifications:

Camshaft Journal Diameter –
Front (No. 1) . . . . 47.47-47.50 mm
(1.869-1.870 inch)
Center (No. 2) . . . . 47.22-47.24 mm
(1.859-1.860 inch)
Rear (No. 3) . . . . 46.71-46.74 mm
(1.839-1.840 inch)

Fig. 71—The camshaft can be withdrawn as shown. The gear can remain attached.

Journal to Bore Clearance –
Front With Bushing . . . . 0.05-0.13 mm
(0.002-0.005 inch)
Front No Bushing . . . . 0.10-0.20 mm
(0.004-0.008 inch)
Center and Rear . . . . 0.10-0.20 mm
(0.004-0.000 inch)

Cam lift should be 7.90-7.98 mm (0.311-0.314 inch). Camshaft journals should not be worn more than 0.05 mm (0.002 inch). Camshaft end play is controlled by the leaf type spring riveted to the rear face of the timing gear cover.

Camshaft bushing in block for the front camshaft journal is renewable. When installing new bushing, face of bushing marked "FRONT" must be forward and oil passage in bushing must be aligned with oil passage in block.

Reverse removal procedure when reinstalling. If camshaft gear was removed, be sure stamped letters "D" shown at (X—Fig. 68) on gear and camshaft flange are aligned. Tighten camshaft gear retaining cap screws to 28 N·m (21 ft.-lbs.) torque.

ROD AND PISTON UNITS

Three-Cylinder Models

54. Connecting rod and piston units are removed from above after removing the cylinder head as outlined in paragraph 37 and the oil pan. Remove carbon and wear ridge from top of cylinder bore, unbolt and remove bearing cap, then push rod and piston assembly out of cylinder. Make sure that cylinder number (C—Fig. 72) is stamped on both rod and cap on side away from camshaft. The side of rod and cap that are marked with cylinder numbers is also the side which has tangs (T) of bearing inserts.

Nuts retaining the connecting rod cap should not be reinstalled. Always install new nuts when reassembling. Two different nuts have been used, but only the latest type should be installed. Early models were originally fitted with cadmium plated (bright finish) nuts while later nuts are phosphated and have a dull black finish. Use only the latest (phosphated) nuts tightened to 81 N·m (60 ft.-lbs.) torque. Early nuts can be tightened to only 61 N·m (45 ft.-lbs.) torque.

It is important that the top of piston is 0.03-0.15 mm (0.001-0.006 inch) below the top face of cylinder block when assembled. Pistons are available with top surfaces different distances from piston pin and also "un-topped," which require machining the piston top to the correct height. After machining top of piston, top inner edge of the combustion chamber in piston should be rounded to a radius of 0.4-0.8 mm (1/64-1/32 inch) and top of piston should be stamped with marks (A and B—Fig. 72) indicating cylinder number and front.

Fig. 72—Assemble rod and piston as shown. "FRONT" or "F" mark (A) must be toward front of engine, numbers indicating cylinder location (C) on rod and cap and tangs (T) on bearing inserts must be together and opposite (away from) camshaft side of engine. Number (B) on top of piston, indicating cylinder location, should be same number as (C) on connecting rod and cap. Cavity in top of piston should be offset toward same side as cylinder numbers (C) on connecting rod and cap.

PISTONS, RINGS AND SLEEVES

Three-Cylinder Models

55. The aluminum alloy, cam ground pistons are supplied in standard size only and are available in a kit consisting of piston, pin and rings for one cylinder. The combustion chamber is offset in piston toward injection nozzle side of engine. Piston is marked "FRONT" or "F" to facilitate assembly. Additionally, the offset combustion chamber should be toward same side as numbers stamped on connecting rod and cap, which should be installed in engine away from camshaft side.

It is important that the top of piston is 0.03-0.15 mm (0.001-0.006 inch) below the top face of cylinder block when assembled. Pistons are available with top surfaces different distances from piston pin and also "un-topped," which require machining the piston top

Fig. 73—Drawing of four-ring piston typical of type used on naturally aspirated engine showing cross section of rings correctly installed. Refer to text.

Fig. 74—Drawing of three-ring piston typical of type used on turbocharged engine showing cross section of rings correctly installed. Refer to text.

to the correct height. Install the connecting rod and piston assembly to the crankshaft and in cylinder block, but without piston rings, mark the top edge of cylinder block with piston at TDC, then remove and machine top of piston as required. After machining top of piston, top inner edge of the combustion chamber in piston should be rounded to a radius of 0.4-0.8 mm ($\frac{1}{64}$-$\frac{1}{32}$ inch) and top of piston should be stamped with marks (A and B—Fig. 72) indicating cylinder number and front.

Four piston rings (Fig. 73) are used on naturally aspirated engines. The first (top) ring is chrome plated and if marked, should be installed with "T" or "TOP" side up. The second and third rings are alike and are internally stepped. The internal groove should be installed toward top of piston. The bottom (fourth), oil control ring is non-directional with a coil spring expander. Install the expander in lower groove first and join ends with the locating wire. Install the ring over the expander, making certain that joint of expander is 180° from (opposite) gap of ring. Piston and ring specifications are as follows:

Four-Ring Pistons

Ring Width –
  Top . . . . 2.36-2.38 mm
                        (0.0928-0.0938 inch)
  Second and Third. . . . . 2.35-2.38 mm
                        (0.0927-0.0937 inch)
  Bottom . . . . 4.72-4.76 mm
                        (0.1862-0.1975 inch)
Ring Side Clearance in Groove –
  Top . . . . 0.05-0.10 mm
                        (0.0019-0.0039 inch)
  Second and Third. . . . . 0.05-0.10 mm
                        (0.002-0.004 inch)
  Bottom . . . . 0.05-0.11 mm
                        (0.002-0.0045 inch)
Ring End Gap –
  Top . . . . 0.36-0.69 mm
                        (0.014-0.027 inch)
  Second and Third. . . . . 0.28-0.61 mm
                        (0.011-0.024 inch)
  Bottom . . . . 0.25-0.74 mm
                        (0.010-0.029 inch)

Three piston rings (Fig. 74) are used on turbocharged engines. The first (top) ring is tapered and the face is chrome plated. The second ring has a tapered face and ring must be installed with "T" or "TOP" mark toward top of piston. The bottom (third), oil control ring is non-directional with a coil spring expander. Install the expander in lower groove first and join ends with the locating wire. Install the ring over the expander, making certain that joint of expander is 180° from (opposite) gap of ring. Piston and ring specifications are as follows:

Three-Ring Pistons
Ring Width –
Top . . . . 2.36-2.38 mm
(0.0928-0.0938 inch)
Second . . . . 2.35-2.38 mm
(0.0927-0.0937 inch)
Bottom . . . . 4.72-4.73 mm
(0.1860-0.1865 inch)
Ring Side Clearance in Groove –
Top . . . . Tapered
Second . . . . 0.05-0.10 mm
(0.002-0.004 inch)
Bottom . . . . 0.05-0.10 mm
(0.002-0.004 inch)
Ring End Gap –
Top . . . . 0.30-0.74 mm
(0.012-0.023 inch)
Second . . . . 0.20-0.64 mm
(0.008-0.025 inch)
Bottom . . . . 0.25-0.76 mm
(0.010-0.030 inch)

On all models, original cylinder bore diameter is 91.48-91.50 mm (3.6015-3.6025 inches). Install new liner if taper, out-of-round or wear exceeds 0.18 mm (0.007 inch). Suitable sleeve removal and installing tools are required to renew sleeves. Thoroughly clean and inspect cylinder block bores and new sleeves before installing, because even the slightest burr or dirt can cause distortion of new sleeve as it is pressed into block. Measure height of liner (sleeve) flange and depth of block recess. Cylinder liner flange should be from 0.10 mm (0.004 inch) below surface of block to 0.10 (0.004 inch) above block surface. Shims are available which may be installed under flange if necessary to raise sleeve. Make sure that block recess is clean. Standard depth is 3.76-3.86 mm (0.148-0.152 inch).

Lubricate cylinder block bores with new engine oil just prior to installing new sleeves. Flanged, prefinished, service sleeves should fit block bores from 0.025 mm (0.001 inch) loose to 0.025 mm (0.001 inch) tight. Correct installed diameter of prefinished service liners is 91.50-91.53 mm (3.6025-3.6035 inches). Allow sufficient time for cylinder liner to stabilize, then check installed diameter of sleeves carefully at several locations from top to bottom of bore and in several different locations around bore.

Unfinished cylinder liners similar to those used in original production are available for service. Unfinished liners are 0.03-0.08 mm (0.001-0.003 inch) interference fit in cylinder block bores and require heavy pressing equipment. Accurate boring equip-

ment should be used to machine installed sleeve to original cylinder bore diameter of 91.48-91.50 mm (3.6015-3.6025 inches). Sleeves with oversize outside diameter are also available for service. Standard block bore is 93.66-93.69 mm (3.6875-3.6885 inches).

PISTON PINS

Three-Cylinder Models

56. The full floating, 31.744-31.750 mm (1.24975-1.25000 inches) diameter piston pin is retained in piston bosses by snap rings. The piston pin is a transitional fit in piston and should have 0.01-0.04 mm (0.0005-0.00175 inch) clearance in bushing at top of connecting rod. To remove piston pin, remove snap rings from both ends, then heat the piston to 40-50° C (100-120° F). Pin should slide out after heating. Do not attempt to press pin out before heating piston.

Bushing in top of rod is renewable. Be sure that oil hole (H—Fig. 72) in bushing is aligned with oil hole in top of rod bore. Inside diameter of new bushing is 31.76-31.79 mm (1.2505-1.2515 inch), providing 0.01-0.04 mm (0.0005-0.00175 inch) clearance for pin. Refer to paragraph 57 for installation of new pin bushing.

CONNECTING RODS AND BEARINGS

Three-Cylinder Models

57. Connecting rod crankpin bearings are steel backed, aluminum/tin faced type renewable from below after removing the oil pan and bearing caps. Projections (tangs) on bearing insert should engage milled slots in rod and cap. The tangs (T—Fig. 72) on bearing inserts must both be located on the same side and away from camshaft. If new connecting rod is installed, cylinder number (C) should be stamped on rod and cap.

Bushing in top of rod for piston pin is renewable. Be sure that oil hole (H) in bushing is aligned with oil hole in top of rod bore while pressing new bushing into position. Size new bushing to 31.76-31.79 mm (1.2505-1.2515 inch) after installing to provide 0.01-0.04 mm (0.0005-0.00175 inch) clearance for pin. Pin bushing must also be parallel to bore for crankshaft crankpin. Connecting rod length between centers is 223.80-223.85 mm (8.811-8.813 inches).

Crankshaft crankpin diameter is 57.11-57.13 mm (2.2484-2.2492 inches) new. Bearing insert are available in standard size and 0.25, 0.50 and 0.75 mm (0.010, 0.020 and 0.030 inch) undersizes. If crankpin is resized, crankpin fillet radii should be 3.97-4.37 mm (0.1562-0.1719 inch). Rod bearing diametral clearance on crankpin should be 0.06-0.10 mm

(0.0023-0.0041 inch) and rod should have 0.24-0.50 mm (0.0095-0.0198 inch) end play on crankpin.

CRANKSHAFT AND BEARINGS

Three-Cylinder Models

58. The crankshaft is supported in four steel backed, aluminum/tin faced type, renewable bearing inserts. Projections (tangs) on bearing insert must engage milled slots in block and main bearing cap. Tangs of inserts in block and cap must both be on same side of engine.

To remove the crankshaft, first drain coolant and engine oil, then remove engine as outlined in paragraph 36. Remove oil pan, timing gear cover, lower cover from timing gear case and engine oil pump. Remove clutch, flywheel, engine adapter plate and crankshaft rear oil seal. Remove connecting rod caps and main bearing caps, then lift crankshaft from cylinder block.

The upper and lower halves of the front main bearing insert are not interchangeable with other bearings or with each other. The insert halves of second and third main bearings are alike, but not interchangeable with other bearing inserts. Main bearing inserts are available in standard size and 0.25, 0.50 and 0.75 mm (0.010, 0.020 and 0.030 inch) undersizes. If crankshaft main journal is resized, journal fillet radii should be 2.38-2.78 mm (0.0937-0.1093 inch).

Crankshaft end play is controlled by renewable thrust washers at front and rear of rear main bearing. The cap half is prevented from turning by the tab which fits in a machined notch of cap. Block half of washer can be rolled from position when cap is removed. Recommended crankshaft end play is 0.05-0.38 mm (0.002-0.015 inch). Thrust washers are available in 0.18 mm (0.007 inch) oversize as well as standard thickness. One set (top and bottom) of oversize thrust washers may be installed on one side (front or rear) in combination with a standard size set on other side to provide oversize adjustment of 0.18 mm (0.007 inch). Two sets of oversize thrust washers may be used to provide adjustment of 0.36 mm (0.014 inch).

Crankshaft balance weights are available only in matched pairs because variation between weights must be less than 28.3 grams (1 ounce). Tighten weight retaining screws to 75 N·m (55 ft.-lbs.) torque, then bend lock tabs to lock screws in place.

Check crankshaft against the values which follow:
Main Journal Standard Diameter . 69.81-69.83 mm (2.7485-2.7493 inches)
Main Journal Fillet Radii . . . . 2.38-2.78 mm (0.0937-0.1093 inch)
Main Bearing Clearance . . . . 0.04-0.10 mm (0.0017-0.0040 inch)
Crankshaft End Play. . . . . 0.05-0.38 mm (0.002-0.015 inch)
Crankpin Standard Diameter. . . . . 57.11-57.13 mm (2.2484-2.2492 inches)
Crankpin Fillet Radii . . . . 3.97-4.37 mm (0.1562-0.1719 inch)
Rod Bearing Clearance on Crankpin . 0.06-0.10 mm (0.0023-0.0041 inch)
Rod End Play on Crankpin . . . . 0.24-0.50 mm (0.0095-0.0198 inch)

To reinstall crankshaft, reverse removal procedure. Notice that main bearing caps are numbered 1 through 4 with number 1 cap at front of engine. The caps are also marked with the serial number of the engine. Make certain thrust washers are installed with grooved side toward crankshaft. Tighten main bearing cap screws to 150 N·m (110 ft.-lbs.) torque.

CRANKSHAFT REAR OIL SEAL

Three-Cylinder Models

59. The lip type crankshaft rear oil seal is contained in retainer at rear of cylinder block. Before installing new seal, inspect crankshaft flange for wear in the seal contact area. Seal can be relocated slightly if necessary so that lip contacts unworn area.

Make sure that seal is pressed squarely into retainer. Lubricate seal lip and crankshaft flange prior to installing seal.

FLYWHEEL

Three-Cylinder Models

60. To remove the flywheel, first separate the engine from transmission housing as outlined in paragraph 152 and remove the clutch. The flywheel is secured to the crankshaft flange by six evenly spaced cap screws. To properly time the flywheel to engine during installation, align unused hole in flywheel with untapped hole in crankshaft flange.

CAUTION: Flywheel is only lightly piloted to crankshaft. Use caution when unbolting to prevent flywheel from falling and possibly causing injury.

The starter ring gear can be renewed after the flywheel is removed. Heat new ring gear to 245° C (475° F) before installing in position. Do not overheat ring gear and heat gear evenly. Heat treatment of gear could be disturbed by improper or excessive heating. Install gear on flywheel with beveled teeth toward front. Allow gear to air cool. Refer to Fig. 146 and paragraph 154 or paragraph 156 if clutch surface of flywheel needs resurfacing.

Reinstall flywheel and tighten retaining cap screws to 106 N·m (78 ft.-lbs.) torque. Check flywheel runout with a dial indicator after flywheel is installed. Runout measurement at machined outer surface should not exceed 0.30 mm (0.012 inch). Runout measured at clutch surface must not exceed 0.025 mm (0.001 inch) for each 25 mm (1 inch) from flywheel center line to the point of dial indicator measurement. If runout is excessive, remove flywheel and check for burrs or foreign material.

OIL PAN

Three-Cylinder Models

61. The cast iron oil pan serves as part of the tractor frame and as attaching point for the tractor front support. To remove the oil pan, first drain engine oil and refer to paragraph 35 to split the front end away from the engine. Support oil pan, remove the attaching screws, then lower oil pan away from the engine.

Reinstall oil pan by reversing removal procedures.

OIL PUMP

Three-Cylinder Models

62. The rotary type oil pump (12—Fig. 75) is mounted on front main bearing cap and is driven from the crankshaft timing gear through an idler. To remove the pump, first remove the oil pan as outlined in paragraph 61. Disconnect the oil delivery pipe from the cylinder block and oil suction pipe from the main bearing cap. Unbolt and remove the front main bearing cap with the oil pump attached. Be careful not to damage the main bearing inserts. The main bearing upper insert may release and fall later if not secured. Pump can be unbolted from main bearing cap after removing the idler gear.

The pump idler gear is retained on shaft by a snap ring, and idler gear end play on shaft should be 0.20-0.58 mm (0.008-0.023 inch). Idler gear is fitted with a bushing which should be renewed if worn excessively. Idler shaft diameter should be 16.63-16.64 mm (0.6548-0.6553 inch). Idler gear bushing should have 0.02-0.06 mm (0.0009-0.0024 inch) clearance on shaft.

Cover can be removed from pump and idler gear can be removed, but drive gear and inner rotor should not be removed from shaft. Clearances can be measured without further disassembly. Check clearance between tip of inner rotor and the outer rotor with a feeler gauge. Rotor tip clearance should be 0.01-0.06 mm (0.0005-0.0025 inch). Check clearance between the outer surface of the outer rotor and bore of pump body with a feeler gauge. Rotor to body clearance should be 0.28-0.33 mm (0.011-0.013 inch). Use a straightedge and feeler gauge to check end clearance of the inner and outer rotors separately. End clearance of inner rotor in pump body should be 0.04-0.08 mm (0.0015-0.0030 inch) and end clearance of outer rotor in body should be 0.01-0.06 mm (0.0005-0.0025 inch). If pump is worn excessively or is excessively scored, the complete pump should be renewed.

Fig. 75—Exploded view of oil pump, lines and oil pan.

1. Oil pan
2. Dipstick
3. Oil pickup
4. Pressure (delivery) line
5. Snap ring
6. Idler gear
7. Bushing
8. Cotter pin
9. Retainer
10. Relief valve spring
11. Relief valve plunger
12. Oil pump

Fig. 76—Block radiator up as shown to remove radiator from tractors with three-cylinder engines.

Reinstall pump by reversing removal procedure. Tighten main bearing retaining screws to 150 N•m (110 ft.-lbs.) torque.

RELIEF VALVE

Three-Cylinder Models

63. The oil pump relief valve is contained in the oil pump body. Remove cotter pin (8—Fig. 75) to release spring (10) and plunger(11). Inspect plunger (11) and bore in pump housing for wear or scoring. Bore diameter is 14.24-14.29 mm (0.5605-0.5625 inch) and plunger diameter is 14.19-14.21 mm (0.5585-0.5595 inch). Spring (10) should exert 34.52-36.66 N (3.52-3.74 kg or 7.76-8.24 lbs.) when compressed to 32.5 mm (1.28 inches). Oil pressure should be 345-445 kPa (50-65 psi) at high idle speed with engine at normal operating temperature.

COOLING SYSTEM

Three-Cylinder Models

64. RADIATOR. Remove the precleaner, hood and both lower panels. Disconnect the headlight wires and remove grille. Remove battery and drain coolant. Block the left side of fuel tank (1—Fig. 76) up so that it will not interfere with removal of the radiator. Disconnect inlet hose from air cleaner and both coolant hoses from radiator. Remove the upper section of exhaust pipe, then unbolt and remove the cross member (2) supporting the front of fuel tank. Unbolt and remove the side rail (3 and 4), then remove the

Fig. 77—Exploded view of water pump and thermostat housing for three-cylinder engines.

1. Pulley & flange	11. Elbow
2. Shaft & bearing	12. Bypass hose
5. Housing	13. Outlet hose
6. Seal assembly	14. Thermostat housing
7. Impeller	15. Gasket
8. Gasket	16. Gasket
9. Back cover plate	17. Thermostat
10. Gasket	18. Elbow

radiator top stay. Disconnect or remove the hydraulic pipe (5) and filter to the rear of tractor. Unbolt fan shroud from radiator and slide shroud toward rear over the fan and coolant pump. Remove the two nuts from mounts at lower front corners of the radiator, then remove radiator from left side.

To reinstall, reverse removal procedure. Coolant capacity is approximately 9.8 L (10.4 qts.) for models without cab. Radiator cap is 0.75 bar (10 psi) rating.

65. THERMOSTAT. The thermostat (17—Fig. 77) is contained in a housing (14) under the coolant elbow (18). The thermostat should begin to open at approximately 81-84° C (177-183° F).

66. COOLANT (WATER) PUMP. To remove the coolant pump, first drain the cooling system, then remove the radiator as outlined in paragraph 64. Remove the radiator shroud, loosen drive belt, then unbolt and remove the fan and belt. Disconnect hoses from pump, remove pump mounting screws, then remove pump.

To disassemble pump, remove cover (9—Fig. 77), then use a suitable puller to remove the pulley and

Fig. 78—Cross section of water pump showing distances to be measured when pressing pump together. Refer to text.

flange (1). Press shaft and bearing assembly (2), together with impeller (7), toward rear out of housing (5).

Inspect all parts for wear or damage and renew as necessary. When reassembling, shorter end of shaft must be toward front. Press shaft and bearing (2) into housing until bearing is flush with front of housing. Support front end of shaft, then press impeller (7) onto shaft until clearance (A—Fig. 78) between front of impeller and inside of pump body is 0.25-0.51 mm

(0.010-0.020 inch). Install pump mounting screws in holes in housing, before installing pulley and flange (1—Fig. 77). Support rear end of shaft (not impeller) then press pulley and flange (2) onto shaft until distance (B—Fig. 78) is 140.49 mm (5.531 inches). Make certain that shaft rotates smoothly.

Reinstall coolant pump by reversing the removal procedure. Adjust fan belt tension to provide 10 mm (3/8 inch) deflection midway between pulleys.

SIX-CYLINDER 6.3544 ENGINE AND COMPONENTS

Six-cylinder 6.3544 engines are installed in early 399 models. Refer to paragraph 35 and following for service to the three-cylinder engines installed in 340, 350, 355 and 360 models. Refer to paragraph 96 and following for service to six-cylinder (1006.6) engines used in late 399 models. Refer to Condensed Service Data for original application. The engine serial number consists of up to 13 letters and numbers located as shown in Fig. 79. The numbers and letters constituting the engine serial number can be used to identify the engine family, country of manufacture and the year of manufacture. Refer to Fig. 80 for identifying code.

FRONT END/ENGINE SPLIT

Early 399 Models With 6.3544 Engine

67. Block both rear wheels to prevent rolling, remove any front mounted equipment and any front weights. Remove hood, grille and side panels. Drain coolant from radiator and engine block. Remove the air cleaner and exhaust. Disconnect battery ground and front wiring harness to the headlights, horn and fuel tank. If necessary, drain fuel tank, then disconnect fuel feed and return lines. Disconnect top and

Fig. 79—View of 354 cubic inch displacement, six-cylinder engine showing location of the engine serial number.

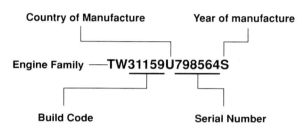

Fig. 80—The first two letters of the engine serial number identifies engine family, the second group is the build code. The letter approximately in the center identifies the country of manufacture and is followed by the engine serial number. The last letter indicates the year of manufacture. The illustration shown above identifies a 6.3544 engine (naturally aspirated, six-cylinder engine with 354 cubic inch displacement), built in the United Kingdom in 1988. The build code is 31159 and the serial number is 798564.

Engine family –
TW = 6.3544 (Turbocharged, 6 cyl., 354 cid)

Country of manufacture –

A = Argentina	P = Poland
B = Brazil	S = India
C = Australia	T = Turkey
D = Germany	U = United Kingdom
E = Spain	W = Iran
F = France	X = Peru
G = Greece	Y = Yugoslavia
J = Japan	**Year of manufacture –**
K = Korea	N = 1986
L = Italy	P = 1987
M = Mexico	S = 1988
N = U.S.A.	T = 1989
	U = 1990
	V = 1991
	W = 1992

bottom radiator hoses, hydraulic pipes to and from the oil cooler and hoses to the steering cylinders. On four-wheel-drive tractors, refer to paragraph 9 and remove drive shaft and shield. On all models, wedge between the front support and axle to prevent tipping and support the front axle, radiator and nose from above using a movable hoist. Support rear of tractor under transmission housing, then carefully remove stud nuts and screws attaching front support to the engine. Carefully roll the front axle, radiator and front assembly forward, away from engine.

Reassemble by reversing the splitting procedure. Refer to paragraph 5 for attaching front support to engine and selecting proper shims. Refer to paragraph 9 for installing drive shaft of four-wheel-drive models.

R & R ENGINE

Early 399 Models
With 6.3544 Engine

68. To remove the engine, first refer to paragraph 67 and split tractor between front support and front of engine. Disconnect wires from engine components and position wires out of the way. Disconnect throttle and stop cables from the fuel injection pump, detach cables from support brackets and move the cable out of the way. Disconnect hydraulic lines between rear of tractor and front that would interfere with the removal of engine and cover all openings. On models with cab, disconnect heater and air conditioning lines if so equipped. Remove gear lever trim plate and boot from models with cab. On all models, attach hoist to the engine, then remove the screws and nuts attaching the engine to transmission housing. Carefully move the engine forward away from the transmission.

To reinstall engine, reverse the removal procedure and observe the following. Install a M12 alignment stud approximately 100 mm (4 inches) long in each side of the transmission housing to assist in aligning engine to transmission. Turn flywheel to align clutch plate splines with transmission and pto input shaft splines. Install retaining screws and nuts after engine and transmission housing flanges are completely together. Tighten retaining screws and nuts to 115 N·m (85 ft.-lbs.) torque. Reverse removal procedure and refer to paragraph 5 for attaching front support to engine and selecting proper shims. Check and adjust clutch linkage as outlined in paragraph 151. Refer to paragraph 9 for installing drive shaft of four-wheel-drive models.

CYLINDER HEAD

Early 399 Models
With 6.3544 Engine

69. REMOVE AND REINSTALL. To remove cylinder head, first drain cooling system, remove hood and disconnect battery ground cable. Disconnect exhaust pipe and inlet pipe from manifolds and top radiator hose. Unbolt and remove inlet manifold, exhaust manifold and coolant outlet housing from cylinder head. Remove the fuel pressure and return

Fig. 81—On 6.3544 engines, two different lengths of studs and three different lengths of screws are used to retain the cylinder head. Refer to text for proper tightening.

lines, then unbolt and remove fuel injector assemblies from cylinder head. Disconnect breather pipe, remove rocker arm cover, then unbolt and remove the rocker arm assembly. Loosen screws retaining cylinder head in reverse of order shown in Fig. 81. Remove screws, then lift cylinder head from tractor. **Do not use sharp or hard tools between cylinder block and head, and lay removed head on clean flat surface to prevent damage.**

Refer to specific paragraphs for servicing rocker arms, valves, seats and guides. Clean gasket surface of cylinder head and check for distortion or other damage. Check flatness of cylinder head gasket surface using a straightedge and feeler gauge. Cylinder head should be renewed or resurfaced if surface is warped more than 0.13 mm (0.005 inch) across head or 0.25 mm (0.010 inch) from front to rear. No more than 0.30 mm (0.012 inch) can be removed from head, and nozzle protrusion should be less than 3.45 mm (0.136 inch). Standard depth of cylinder head is 94.87-95.63 mm (3.735-3.765 inches).

Different head gaskets and fasteners have been used and are available for service. Refer to the following tightening information for the specific engine model, fastener type and gasket used.

On **early 399 models with 6.3544 engine,** cylinder head gasket should be marked "TOP FRONT" and should be installed dry. The cylinder head retaining fasteners may have separate washers or may have washers integral with the nuts or screws. Only one type of fastener (with integral washers or separate washers) should be used. Tighten fasteners in sequence shown in Fig. 81. If fasteners have separate washers, tighten fasteners (1 through 32) cold to 130 N•m (96 ft.-lbs.) torque and smaller fasteners (33 through 38) to 38 N•m (28 ft.-lbs.) torque. Retorque retaining screws (1 through 32) hot to 142 N•m (105 ft.-lbs.) after running engine until warm and again after operating for 25-50 hours. Nuts retaining injection nozzles should be tightened to 16 N•m (12 ft.-lbs.) torque.

The cylinder head can be installed and the fasteners tightened only once (no retorque) if late fasteners with integral washers are used. Fasteners should be new or in very good condition if the "no retorque" method is used. Check all screws and studs for distortion and elongation using a straightedge along

side of the threads. Discard any screw or stud if any part appears to have a reduction in diameter. Install the cylinder head gasket and lubricate the threads of fasteners. Install stud nuts and screws, then gradually and evenly tighten fasteners indicated (1 through 32—Fig. 81) to 100 N•m (74 ft.-lbs.) torque using the sequence shown. Check to be sure that all are tightened to exactly 100 N•m (74 ft.-lbs.) torque, then tighten each fitting exactly the number of degrees specified for its length as follows. Short screws (5, 6, 11, 12, 16, 17, 21, 22, 26, 27, 31 and 32) should be turned 150° (2½ flats). Longer screws (2, 8, 13, 18, 23 and 28) and stud nuts (1, 3, 4, 7, 9, 10, 14, 15, 19 and 20) should be turned 180° (3 flats or ½ turn). Stud nuts on the longer studs (24, 25, 29 and 30) should be turned 210° (3½ flats). Tighten the smaller 5/16 inch screws (33 through 38—Fig. 81) to 38 N•m (28 ft.-lbs.) torque. The following lists, in order, the number of degrees (and number of flats) to tighten screws 1 through 32.

Fastener Angle

1-4	180° (3 flats)
5 and 6	150° (2½ flats)
7-10	180° (3 flats)
11 and 12	150° (2½ flats)
13-15	180° (3 flats)
16 and 17	150° (2½ flats)
18-20	180° (3 flats)
21 and 22	150° (2½ flats)
23	180° (3 flats)
24 and 25	210° (3½ flats)
26 and 27	150° (2½ flats)
28	180° (3 flats)
29 and 30	210° (3½ flats)
31 and 32	150° (2½ flats)

If screws and studs are in good condition and if tightening is carefully and accurately done, the fasteners will retain the correct torque and will not require retightening. Special tools are available to assist in turning the fasteners the specified number of degrees, but scribing match marks on the fastener and the cylinder head before turning will also be helpful. Nuts retaining injection nozzles should be tightened to 16 N•m (12 ft.-lbs.) torque.

VALVES AND SEATS

Early 399 Models With 6.3544 Engine

70. When disassembling the cylinder head, keep valves in order so that they can be reinstalled in original location, if reused. Valve heads are marked with cylinder numbers consecutively from front to rear on some engines. Any replacement valves should be similarly marked before installation.

Inlet and exhaust valves originally seat directly in cylinder head; however, cylinder head may be machined and fitted with replacement seats. Inlet and exhaust valve face angles are 45° and seat angles are 46°. Inlet and exhaust valves should be recessed 1.02-1.27 mm (0.040-0.050 inch) below surface of cylinder head and should not be recessed more than 1.52 mm (0.060 inch). Clearance between valve stem and guide should not exceed 0.13 mm (0.005 inch) for inlet valves; 0.15 mm (0.006 inch) for exhaust valves. Standard valve stem diameter is 9.46-9.49 mm (0.3725-0.3735 inch) for inlet valves; 9.42-9.45 mm (0.3710-0.3720 inch) for exhaust valves. Refer to paragraph 71 for service to valve guides. Cup type valve stem seals are used on all valves.

VALVE GUIDES

Early 399 Models
With 6.3544 Engine

71. Inlet and exhaust valves operate in machined bores of valve guides that are pressed into the cylinder head. Clearance between valve stem and guide should not exceed 0.13 mm (0.005 inch) for inlet valves; 0.15 mm (0.006 inch) for exhaust valves. Standard valve guide inside diameter is 9.51-9.54 mm (0.3743-0.3757 inch). Length of inlet guide is 57.94 mm (2.281 inches); 61.11 mm (2.406 inches) for exhaust guides.

New service valve guides can be installed if original guides are worn. Press guides into cylinder head until top of guide is 15.09 mm (0.594 inch) above spring seating face. Be sure to reface valve seats after renewing valve guides. Cup type valve stem seals are used on all valves.

VALVE SPRINGS

Early 399 Models
With 6.3544 Engine

72. Valve springs, retainers and locks are interchangeable for inlet and exhaust valves. Cup type valve stem seals are used on all valves. Close wound damper coils should be installed toward cylinder head. Renew springs if discolored, distorted or if they fail to meet the following specifications. Outer spring should exert 17.25-19.52 kg (38-43 lbs.) when compressed to 45.20 mm (1.780 inches). Inner spring should exert 6.64-7.57 kg (14.63-16.67 lbs.) when compressed to 39.7 mm (1.563 inches).

CAM FOLLOWERS

Early 399 Models
With 6.3544 Engine

73. The mushroom type cam followers operate directly in the machined bores in cylinder block. Cam followers can be removed through bottom of cylinder block after removing timing gear housing as outlined in paragraph 76, the engine oil pan and camshaft as outlined in paragraph 82. Identify cam followers so they can be reinstalled in their original locations if reused.

The 18.99-19.01 mm (0.7475-0.7485 inch) diameter cam followers are available only in standard size and should have 0.04-0.09 mm (0.0015-0.00375 inch) diametral clearance in block bores. Install new cam followers if scored, pitted or excessively worn.

ROCKER ARMS

Early 399 Models
With 6.3544 Engine

74. The rocker arms and shaft assembly can be removed after removing the hood and rocker arm cover. The rocker arms are right and left hand units and should be assembled as shown in Fig. 82.

Rocker shaft diameter is 19.01-19.04 mm (0.7485-0.7495 inch) and the desired diametral clearance between shaft and rocker arms is 0.03-0.09 mm (0.0010-0.0035 inch). Renew shaft and/or rocker arms if clearance exceeds 0.13 mm (0.005 inch) or if either surface of shaft or bushing in rocker arm is rough.

Oil is supplied to the rocker arm shaft through the oil feed connection located between cylinders two and three on the rocker arm shaft. Two types of oil feed connections have been used. Early engines were fitted with brass oil feed connection without spacers. Later engines and service replacement oil feed connection is zinc and washers are installed between oil feed connection and springs. Make sure that seal ring for base of oil feed connection is in place when installing rocker arm assembly. Tighten rocker shaft support bracket retaining screws to 75 N·m (55 ft.-lbs.) torque. Tighten retaining screws evenly, beginning in the center.

VALVE CLEARANCE

Early 399 Models
With 6.3544 Engine

75. Recommended valve clearance is 0.2 mm (0.008 inch) cold for inlet valves; 0.45 mm (0.018 inch) cold for exhaust valves. Static setting of all valves can be made from two crankshaft positions using the following procedure:

Remove the timing window plug from the left side of flywheel adapter housing and turn the flywheel until "TDC 1" mark on flywheel is aligned in center of hole. Check valves for front cylinder. Both valves should be closed if front cylinder is on compression stroke. If front two valves are tight (open), turn crankshaft exactly one complete revolution to set front piston on compression stroke.

Fig. 82—With TDC timing marks aligned and number 1 (front) piston on compression stroke, adjust the valves indicated. After adjusting these valves, turn crankshaft exactly one revolution (until marks again align) then adjust valves shown in Fig. 83.

Fig. 83—With TDC timing marks aligned and number 1 (front) piston on exhaust stroke, adjust the valves indicated.

Fig. 84—Crankshaft pulley and related parts. Refer to text for recommended removal procedure.

1. Thrust block
2. "O" ring
3. Outer taper ring
4. Inner taper ring
5. Spacer
6. Woodruff key

If not equipped with hole in flywheel adapter or mark on flywheel, turn flywheel in normal direction of rotation until the exhaust valve for the number 6 (rear) cylinder is just about closed and the inlet valve is just beginning to open. If not equipped with marks, mark crankshaft pulley so that crankshaft can be rotated exactly one complete revolution from this position after adjusting the first set of valves.

With front cylinder set at TDC on compression stroke, adjust the six valves indicated in Fig. 82. Turn the crankshaft exactly one complete revolution and adjust the six valves indicated in Fig. 83.

CRANKSHAFT PULLEY

Early 399 Models
With 6.3544 Engine

76. REMOVE AND REINSTALL. To remove the crankshaft pulley, first refer to paragraph 67 and split tractor between front support and front of engine. Loosen alternator and air conditioning compressor (if so equipped), then remove drive belts. Remove the three screws retaining thrust block (1—Fig. 84) to crankshaft, then withdraw thrust block and "O" ring (2). Pulley may be free to remove from crankshaft; however, it may be necessary to bump hub of pulley to separate the tapered locking rings (3 and 4). **Do not use a puller.** Using a puller would cause tapered rings to lock and crankshaft or pulley may be damaged. Remove tapered rings (3 and 4), spacer (5), pulley and key (6).

Install the crankshaft pulley, spacer (5), tapered rings (3 and 4), "O" ring (2) and thrust block (1). Install ring (4), which is cut away on outside before installing ring (3). Install and tighten the three retaining screws to 88 N·m (65 ft.-lbs.) torque.

TIMING GEARS

Early 399 Models
With 6.3544 Engine

77. CAMSHAFT GEAR. Gears are keyed to camshaft and crankshaft and valve timing will be correct if timing marks are properly aligned as shown in Fig. 86. There is no adjustment provided and if valve timing is incorrect, gears are one or more teeth out of correct mesh. The valves can (and probably will) hit the pistons if either the camshaft or crankshaft is turned independently of the other, unless the rocker arms are removed.

Marked teeth of camshaft gear (7) and the idler gear (17) can be viewed, and camshaft gear (7) can be removed and relocated, after removing the cover (1—Fig. 85). Because the idler gear has an odd number of teeth, timing marks on idler gear will only occasion-

Fig. 85—Covers (1 and 2) and gears (7 and 10) must be removed before timing gear housing (12) can be removed for access to remaining gears. Refer to text for details.

1. Cover	9. Retaining plate	16. Washer	24. Injection pump gear
2. Cover	10. Accessory shaft	17. Right idler	25. Oil spray tube
3. Screw	drive gear	18. Bushing	26. Accessory drive shaft
4. Washer	11. Oil seal	19. Hub	27. Bushing
5. Tab washer	12. Timing gear housing	20. Left idler	28. Seal ring
6. Washer	13. Washer	21. Thrust washer	29. Seal
7. Camshaft gear	14. Pin	22. Bushing	30. Seal ring
8. Screw	15. Camshaft	23. Thrust collar	31. Thrust sleeve

ally align with marks on crankshaft and camshaft gears.

To check gear timing, the timing gear housing must be removed as outlined in paragraph 79.

Camshaft gear mounting screw (3—Fig. 85) should be tightened to 68 N•m (50 ft.-lbs.) torque and retaining screws (8) should be tightened to 30 N•m (22 ft.-lbs.) torque.

The injection pump drive must also be timed as outlined in paragraph 78.

78. INJECTION PUMP/ACCESSORY DRIVE GEAR. Turn the crankshaft until the "TDC 1" timing mark on flywheel is aligned as shown in Fig. 88. Front (number 1) cylinder should be on compression stroke. Refer to paragraph 143 and remove the fuel injection pump. Pump timing marks should be aligned as shown in Fig. 91 if front cylinder is on compression stroke. If mark on pump drive gear and master spline

Fig. 86—Front view of engine block with timing marks properly aligned. Timing gear train will never appear as shown because gears (7 and 10) must be removed before timing gear housing. The fuel injection pump, which is driven by accessory shaft drive gear (10), is timed independently.

CR. Crankshaft gear
7. Camshaft gear	17. Right idler
10. Accessory shaft gear	20. Left idler

(S) are located 180° from location shown in Fig. 91, piston may not be on compression stroke. Turn crankshaft one complete revolution and again align TDC mark as shown in Fig. 88. If engine will not run, it is possible that injection pump timing is 180° out of time. Refer to paragraph 79 and check valve timing, then align pump timing marks while camshaft gear timing mark is in mesh with idler gear.

If pump timing is not correct, gear (10—Fig. 85 or Fig. 89) must be relocated on drive shaft (26). Small adjustments are possible by removing cover (2), loosening the mounting screws (8), turning shaft (26) as required to align marks (N—Fig. 91), then retightening screws (8—Fig. 85 or Fig. 89) to 30 N·m (22 ft.-lbs.) torque.

TIMING GEAR HOUSING

Early 399 Models
With 6.3544 Engine

79. REMOVE AND REINSTALL. The timing gear housing (12—Fig. 85) can only be removed after removing the camshaft gear (7), accessory shaft drive gear (10) and the crankshaft pulley. Before timing gear housing can be reinstalled, it is necessary to remove the injection pump for proper retiming of the accessory shaft drive.

To remove the timing gear housing (12) or remove the crankshaft oil seal (11), first refer to paragraph 67 and split tractor between front support and front of engine. Loosen alternator and air conditioning compressor (if so equipped), then remove drive belts. Remove the rocker arm cover and the rocker arms to avoid possible damage to pistons or valve train.

Turn the crankshaft until the "TDC 1" timing mark on flywheel is aligned as shown in Fig. 88. Refer to paragraph 76 and remove the crankshaft pulley. Remove the two covers (1 and 2—Fig. 85), then remove the two gears (7 and 10). It may be necessary to use a puller to withdraw gear (7) from camshaft. Timing gear housing (12) can be unbolted and removed after gears (7 and 10) are removed.

The crankshaft gear and idler gears are now accessible to inspection, removal or service. Refer to paragraph 81 for service to the accessory drive shaft. Refer to paragraph 82 for service to the engine camshaft.

Press old crankshaft oil seal from timing gear housing, then press new seal in from front with spring loaded lip toward inside. Front of crankshaft oil seal should be 6.73-7.24 mm (0.265-0.285 inch) from front face of housing.

Unbolt and remove the right idler gear (17—Fig. 86), then reinstall with two marked teeth (timing marks) aligned with the marked tooth of the crankshaft gear as shown in Fig. 86. Tighten idler gear retaining nuts to 49 N·m (36 ft.-lbs.) torque. If not

already correctly positioned, turn accessory drive shaft (26—Fig. 85) until injection pump timing marks are aligned as shown in Fig. 91.

Make sure that camshaft (15—Fig. 85) and thrust washer (13) are properly positioned and that accessory drive shaft (26) and thrust washer (21) are correctly positioned, then install timing gear housing using a new gasket. Be careful not to damage crankshaft seal and install retaining screws loosely. Center timing gear cover around the crankshaft using special centering tool (PD163 available from Massey-Ferguson) or equivalent, then tighten the retaining screws. If special centralizing tool is not available, the crankshaft pulley can be used carefully, but damage to the soft lip of the seal is likely.

Make sure that camshaft (15) is properly positioned, then install the camshaft gear (7), aligning the two marks on camshaft gear with the single mark on idler gear (Fig. 86). Install washers (4, 5 and 6—Fig. 85) and tighten screw (3) to 68 N·m (50 ft.-lbs.) torque. Be careful not to move shafts when tightening screw and bend tab of washer against screw (3).

Check to make sure that injection pump timing marks are still aligned as shown in Fig. 91, then install accessory drive gear (10—Fig. 85), plate (9) and retaining screws (8). Tighten retaining screws (8) to 30 N·m (22 ft.-lbs.) torque.

Install the crankshaft pulley, spacer (5), tapered rings (3 and 4) "O" ring (2) and thrust block (1) as outlined in paragraph 76.

CRANKSHAFT GEAR AND IDLER GEARS

Early 399 Models
With 6.3544 Engine

80. R&R AND OVERHAUL. The crankshaft gear and idler gears are accessible to inspection, removal or service after removing the timing gear housing as outlined in paragraph 79. Backlash between any two gears in timing gear train should be at least 0.08 mm (0.003 inch). Diametral clearance between idler gear bushings (18—Fig. 85) and hubs (19) should be 0.05-0.13 mm (0.0020-0.0052 inch). Bushings in idler gears must be reamed after installation to inside diameter 47.63-47.67 mm (1.8750-1.8766 inches). End play of idler gears should be 0.05-0.30 mm (0.002-0.012 inch). Refer to Fig. 87 which shows oil passage in idler gear hub and cylinder block.

The crankshaft gear is keyed in place and fits the shaft with a transitional fit, 0.020 mm (0.0008 inch) tight to 0.025 mm (0.0010 inch) loose. If the old gear is loose on shaft, it may be possible to pry gear from shaft, but a puller may be required. If necessary to remove crankshaft gear with a puller, it is suggested to first remove the crankshaft.

Fig. 87—Pressure oil passage (O) in idler gear hub and block.

Fig. 88—View of flywheel "TDC 1" mark with engine at Top Dead Center.

Fig. 89—View of accessory shaft, gears and injection pump drive.

2. Cover
8. Screw
9. Retaining plate
10. Accessory shaft drive gear
12. Timing gear housing
21. Thrust washer halves
22. Bushing
23. Thrust collar
24. Injection pump gear
25. Oil spray tube
26. Accessory drive shaft
27. Bushing
28. Seal ring
29. Seal
30. Seal ring
31. Thrust sleeve
32. Gasket
33. "O" ring
34. Bushing
35. Pump adapter
36. Gasket
37. Fuel injection pump

Refer to paragraph 81 for service to the accessory drive shaft. Refer to paragraph 82 for service to the engine camshaft.

ACCESSORY/INJECTION PUMP DRIVE

Early 399 Models
With 6.3544 Engine

81. R&R AND OVERHAUL. The accessory drive shaft (26—Fig. 89) drives the engine oil pump, the fuel injection pump and the steering or tandem hydraulic pump. Timing of the fuel injection pump is critical to the operation of the engine. Refer to Fig. 90 for cross section of accessory drive assembly.

Refer to paragraph 78 for checking and adjusting gear timing. To remove the accessory drive shaft, first turn the engine until front cylinder is on compression stroke and the flywheel "TDC 1" mark is in timing window as shown in Fig. 88. Remove the fuel injection pump as outlined in paragraph 143, the steering tandem pump as outlined in paragraph 29 and the timing gear housing as outlined in paragraph 79, then withdraw shaft (26—Fig. 89) with a turning motion to disengage worm gear from pump drive gear (24). To remove the injection pump drive gear (24), unbolt and remove adapter (35), then remove thrust sleeve (31) and gear. Lower thrust collar (23) and bushing (22) can be removed using a suitable puller.

Accessory shaft thrust washer (21) is made in two pieces which fit in a machined slot in shaft (26) and machined recess in front face of engine block. Thrust washer (21) should be 0.4.76-4.84 mm (0.1875-0.1905 inch) thick and should protrude 0.01-0.17 mm (0.0005-0.0070 inch) from block recess. The thrust washer is retained in block recess by the timing gear housing and shaft end play should be limited to the clearance of the thrust washer in shaft groove which should be 0.064-0.230 mm (0.0025-0.0090 inch).

Accessory drive shaft diameter at front bushing journal should be 49.16-49.19 mm (1.9355-1.9365 inches) and clearance should be 0.025-0.110 mm (0.001-0.0042 inch) in front bushing. Diameter of accessory drive shaft at rear bushing journal should be 31.70-31.72 mm (1.248-1.249 inches) and clearance in rear bushing should be 0.025-0.100 mm (0.001-0.0039 inch). Bushings are precision type and will need no reaming if carefully installed. Make sure that oil holes in bushings are aligned with passages in block.

Lower journal diameter of drive gear (24—Fig. 89 or Fig. 90) should be 41.18-41.22 mm (1.6214-1.6224 inches) and clearance in lower bushing should be 0.08-0.13 mm (0.0031-0.0052 inch). Bushing (22) may require reaming to size after pressing into adapter (23). Standard fitted size of lower bushing is 41.29-41.32 mm (1.6255-1.6266 inches).

Upper journal diameter of drive gear (24) should be 47.53-47.57 mm (1.8714-1.8730 inches) and clearance in upper bushing should be 0.05-0.13 mm (0.0020-0.0052 inch). Bushing (34) may require reaming to size after pressing into pump adapter (35). Standard fitted size of upper bushing is 47.63-47.67 mm (1.8750-1.8766 inches).

Sealing ring (30) is 2.36-2.38 mm (0.0928-0.0938 inch) thick and should have 0.05-0.10 mm (0.0019-0.0039 inch) clearance in groove of thrust sleeve (31).

The accessory drive journals and bushings are lubricated by the engine oiling system. Additionally, a renewable spray tube (25) provides a spray of lubricating oil for the gears. The engine spray tube can be

Fig. 90—Cross sectional view of accessory drive housing, injection pump drive and engine oil pump drive. Oil spray tube (25) directs lubricant to worm of accessory shaft (26). Refer to Fig. 89 for parts identification.

Fig. 91—Injection pump drive adapter with pump removed showing timing marks. Refer to text for shaft timing procedure.

M. Pump timing mark   S. Master spline
N. Gear timing notches   35. Adapter

removed when cleaning the engine block after removing passage blocking plug. If service block is not be fitted with spray tube, transfer tube from old block or install new spray tube.

When assembling accessory drive, turn engine crankshaft until "TDC 1" mark is aligned in timing window as shown in Fig. 88, then install drive gear (24—Fig. 89), accessory shaft (26) and related parts in cylinder block. Do not install fuel injection pump until timing gear housing is installed. Turn accessory drive shaft (26) until timing notches (N—Fig. 91) on pump adapter (35) and drive gear are aligned as shown and make sure that thrust washer (21—Fig. 89) is fully in recess of cylinder block, then refer to paragraph 79 and install the timing gear housing. Install washers (4, 5 and 6—Fig. 85) and tighten screw (3) to 68 N•m (50 ft.-lbs.) torque. Be careful not to move shafts when tightening screw (3). Tighten retaining screws (8) to 30 N•m (22 ft.-lbs.) torque.

CAMSHAFT

Early 399 Models
With 6.3544 Engine

82. To remove the camshaft, first remove the rocker arm assembly as outlined in paragraph 74, separate the front axle and support from the front of the engine as described in paragraph 67, then refer to paragraph 79 and remove the timing gear housing. Unbolt and remove the fuel lift pump from the right side of engine block. Remove engine oil pan, push the cam followers up away from cam lobes and withdraw camshaft. Remove cam followers from below and thrust washer (13—Fig. 85) from front as camshaft is withdrawn from block. Identify cam followers for installation in the same location if they are to be reinstalled.

Cam lift is 7.617-7.694 mm (0.2999-0.3029 inch) for inlet lobes; 7.711-7.788 mm (0.3036-0.3066 inch) for exhaust lobes. Refer to paragraph 73 for service to cam followers. The camshaft is supported by four journals which operate directly in unbushed cylinder block bores. Refer to the following for journal diameters and desired clearances.

Journal # 1
Diameter. . . . . 50.71-50.74 mm
(1.9965-1.9975 inches)
Clearance . . . . 0.06-0.11 mm
(0.0025-0.0045 inch)

Journal # 2
Diameter. . . . . 50.46-50.49 mm
(1.9865-1.9875 inches)
Clearance . . . . 0.06-0.14 mm
(0.0025-0.0055 inch)

Journal # 3
Diameter. . . . . 5.020-50.23 mm
(1.9765-1.9775 inches)
Clearance . . . . 0.06-0.14 mm
(0.0025-0.0055 inch)

Journal # 4
Diameter. . . . . 49.95-49.98 mm
(1.9665-1.9675 inches)
Clearance . . . . 0.06-0.14 mm
(0.0025-0.0055 inch)

Camshaft end play is controlled by the thickness of the thrust washer (13—Fig. 85) and the width of slot formed by front shoulder on camshaft and rear face of cam gear (7). Thrust washer should be 5.49-5.54 mm (0.216-0.218 inch) thick and should limit camshaft end play to 0.10-0.41 mm (0.004-0.016 inch). The thrust washer should fit in block recess no more than 0.05 mm (0.002 inch) below flush with front face of block, but should not protrude more than 0.08 mm (0.003 inch) above. When reassembling, be sure that thrust washer fits recess correctly and engages locating pin.

When reassembling, refer to paragraph 79. Timing marks on flywheel (Fig. 88), crankshaft gear (CR—Fig. 86), idler gear (17), camshaft gear (7) and injection pump drive (N—Fig. 91) must all be aligned for proper timing. Install washers (4, 5 and 6—Fig. 85) and tighten screw (3) to 68 N•m (50 ft.-lbs.) torque. Be careful not to move shafts when tightening screw and bend tab of washer against screw (3). Tighten retaining screws (8) to 30 N•m (22 ft.-lbs.) torque.

ROD AND PISTON UNITS

Early 399 Models
With 6.3544 Engine

83. Connecting rod and piston units are removed from above after removing the cylinder head as outlined in paragraph 69 and the engine oil pan as outlined in paragraph 90. Remove carbon and wear ridge from top of cylinder bore, unbolt and remove bearing cap, then push connecting rod and piston assembly out of cylinder. Make sure that cylinder number is stamped on both rod and cap (C—Fig. 92) on side away from camshaft. The side of rod and cap that is marked with cylinder number is also the side which has tangs (T) of bearing inserts.

Notice that two different types of connecting rod nuts have been used and each has different tightening torques. Early models were originally fitted with cadmium plated (bright finish) nuts which should be tightened to only 102 N•m (75 ft.-lbs.) torque. Later nuts are phosphated, have a dull black finish and should be tightened to 129 N•m (95 ft.-lbs.) torque.

It is important that the top of piston is correct distance from the top face of cylinder block when assembled. Pistons are available with top surfaces different distances from piston pin and also "un-topped," which require machining the piston top to the correct height. Refer to paragraph 84.

PISTONS, RINGS AND SLEEVES

Early 399 Models
With 6.3544 Engine

84. The aluminum alloy, cam ground pistons are supplied in standard size only. The combustion chamber is offset in piston toward injection nozzle side of engine. Piston is marked "FRONT" or "F" (B—Fig. 92) to facilitate assembly. Additionally, the offset combustion chamber should be toward same side as numbers stamped on connecting rod and cap, which should be installed in engine away from camshaft side.

It is important that the top of piston is flush to 0.18 mm (0.007 inch) above the top surface of the cylinder block when assembled. Pistons are available with top surfaces different distances from piston pin and also "un-topped," which require machining the piston top to the correct height. Install the connecting rod and piston assembly to the crankshaft and in cylinder block, but without piston rings. Mark the top edge of cylinder block with piston at TDC, then remove and machine top of piston as required. After machining top of piston, top inner edge of the combustion chamber in piston should be rounded slightly and top of piston should be stamped with cylinder number (A) and "F" or "FRONT" (B).

Three piston rings (Fig. 92) are used. The chrome plated first (top) ring is barrel faced and internally chamfered. The internal chamfer of the top ring, side of ring marked "T", "TOP" or with the manufacturer's mark should be installed toward top of piston. The second ring is not stepped or chamfered, but the tapered face is chrome plated. The side of the second ring marked "T", "TOP" or with the manufacturer's mark should be toward top of piston. The bottom (third), oil control ring is non-directional with a coil spring expander. Install the expander in lower groove first and join ends with the locating wire. Install the ring over the expander, making certain that joint of expander is 180° from (opposite) gap of ring. Piston and ring specifications are as follows.

Three-Ring Pistons
Ring Width–
  Top . . . . 2.36-2.37 mm
                                       (0.0930-0.0935 inch)
  Second . . . . 2.36-2.37 mm
                                         (0.0930-0.0935 inch)

  Bottom . . . . 4.72-4.76 mm
                                         (0.1860-0.1875 inch)
Ring Side Clearance in Groove–
  Top . . . . 0.06-0.12 mm
                                         (0.0024-0.0048 inch)
  Second . . . . 0.06-0.10 mm
                                         (0.0024-0.0040 inch)
  Bottom . . . . 0.03-0.08 mm
                                         (0.0010-0.0033 inch)
Ring End Gap–
  Top . . . . 0.25-0.71 mm
                                         (0.010-0.028 inch)
  Second . . . . 0.20-0.61 mm
                                         (0.008-0.024 inch)
  Bottom . . . . 0.25-0.84 mm
                                         (0.010-0.033 inch)

Original cylinder bore diameter is 98.48-98.50 mm (3.877-3.878 inches). Install new liner if taper, out-of-round or wear exceeds 0.2 mm (0.008 inch). Suitable sleeve removal and installing tools are required to renew sleeves. Thoroughly clean and inspect cylinder block bores and new sleeves before installing, because even the slightest burr or dirt can cause distortion of new sleeve as it is pressed into block. Measure height of liner (sleeve) flange and depth of block recess. Cylinder liner flange should be from 0.10 mm (0.004 inch) below surface of block to 0.05 (0.002 inch) above block surface. Shims are available which may be installed under flange if necessary to raise sleeve. Make sure that block recess is clean. Standard depth of block recess is 3.81-3.91 mm (0.150-0.154 inch).

Fig. 92—Drawing of piston and connecting rod correctly assembled with "F" or "FRONT" mark (B), cylinder number (A) and rod correlation numbers (C). Notches in rod and cap for tangs (T) should be on side away from camshaft. Mating joint of connecting rod and cap are serrated.

Lubricate cylinder block bores with new engine oil just prior to installing new sleeves. Flanged, prefinished, service sleeves should fit block bores from 0.025 mm (0.001 inch) loose to 0.025 mm (0.001 inch) tight. Correct installed diameter of prefinished service liners is 98.48-98.54 mm (3.877-3.8795 inches). Allow sufficient time for cylinder liner to stabilize, then check installed diameter of sleeves carefully at several locations from top to bottom of bore and in several different locations across bore.

Unfinished cylinder liners similar to those used in original production are available for service. Unfinished liners are 0.03-0.08 mm (0.001-0.003 inch) interference fit in cylinder block bores and require heavy pressing equipment. Accurate boring equipment should be used to machine installed sleeve to original cylinder bore diameter of 98.48-98.50 mm (3.877-3.878 inches). Sleeves with oversize outside diameter are also available for service. Standard block bore is 103.24-103.26 mm (4.0645-4.0655 inches).

PISTON PINS

Early 399 Models
With 6.3544 Engine

85. The full floating piston pin is retained in piston bosses by snap rings.

Piston pin diameter is 34.92-34.930 mm (1.3748-1.3750 inches). The piston pin should have 0.003-0.014 mm (0.00012-0.00055 inch) clearance in piston and should have 0.020-0.043 mm (0.0008-0.0017 inch) clearance in bushing at top of connecting rod. To remove piston pin, remove snap rings from both ends, then heat the piston to 40-50° C (100-120° F). Pin should slide out after heating.

Bushing in top of rod is renewable. Be sure that oil hole (H—Fig. 92) in bushing is aligned with oil hole in top of rod bore. Bushing must be reamed to provide correct clearance for pin. Refer to paragraph 86 for installation of new pin bushing.

CONNECTING RODS AND BEARINGS

Early 399 Models
With 6.3544 Engine

86. Connecting rod crankpin bearings are steel backed, aluminum/silicone or aluminum/tin faced type, renewable from below after removing the oil pan and rod bearing caps. Projections (tangs) on bearing insert should engage milled slots in rod and cap. The tangs (T—Fig. 92) on bearing inserts must both be on same side of engine and away from camshaft. If new connecting rod is installed, cylinder number (C) should be stamped on rod and cap.

Bushing in top of rod for piston pin is renewable. Be sure that oil hole in bushing is aligned with oil hole in top of rod bore while pressing new bushing into position. Size new, installed bushing to 34.94-34.96 mm (1.37575-1.37650 inches). Clearance between piston pin and correctly sized bushing should be 0.020-0.043 mm (0.0008-0.0017 inch). Pin bushing must also be parallel to bore for crankshaft crankpin. Connecting rod length between centers is 219.05-219.10 mm (8.624-8.626 inches).

Crankshaft crankpin diameter is 63.47-63.49 mm (2.4988-2.4996 inches) new. Bearing insert are available in standard size and 0.25, 0.50 and 0.75 mm (0.010, 0.020 and 0.030 inch) undersizes. If crankpin is resized, crankpin fillet radii should be 3.68-3.96 mm (0.145-0.156 inch). Rod bearing diametral clearance on crankpin should be 0.03-0.08 mm (0.0012-0.0031 inch).

CRANKSHAFT AND BEARINGS

Early 399 Models
With 6.3544 Engine

87. The crankshaft is supported in seven steel backed, aluminum/silicone faced type, renewable bearing inserts. Projections (tangs) on bearing insert must engage milled slots in block and main bearing cap. Tangs of inserts in block and cap must both be on same side of engine.

To remove the crankshaft, first drain coolant and engine oil, then remove engine as outlined in paragraph 68. Remove oil pan, then refer to paragraph 79 and remove timing gear housing. Remove clutch, flywheel, engine adapter plate and crankshaft rear oil seal. Remove connecting rod caps, main bearing caps, front and rear block bridge pieces, then lift crankshaft from cylinder block.

The upper and lower halves of main bearing inserts are not interchangeable. The upper insert halves are slotted to provide pressure lubrication to crankshaft and connecting rods. Inserts are interchangeable in pairs (upper and lower halves) for all journals except the center main bearing. Main bearing inserts are available in standard size and 0.25, 0.50 and 0.75 mm (0.010, 0.020 and 0.030 inch) undersizes. If crankshaft main journal is resized, journal fillet radii should be 3.68-3.96 mm (0.145-0.156 inch).

Crankshaft end play is controlled by renewable thrust washers at front and rear of center main bearing. The cap half is prevented from turning by the tab which fits in a machined notch of cap. Block half of washer can be rolled from position when cap is removed. Recommended crankshaft end play is 0.05-0.38 mm (0.002-0.015 inch). Thrust washers are available in 0.18 mm (0.007 inch) oversize as well as standard thickness. One set (top and bottom) of oversize thrust washers may be installed on one side (front or rear) in combination with a standard size set on other side to provide oversize adjustment of 0.18 mm (0.007 inch). Two sets of oversize thrust washers may be used to provide adjustment of 0.36 mm (0.014 inch).

Check crankshaft against the values which follow:

Main Journal
  Standard Diameter . . . . 76.16-76.19 mm
                            (2.9984-2.9996 inches)
Main Journal Fillet Radii . . . . 3.68-3.96 mm
                            (0.145-0.156 inch)
Main Bearing Clearance . . . . 0.046-0.107 mm
                            (0.0018-0.0042 inch)
Crankshaft End Play . . . . 0.05-0.38 mm
                            (0.002-0.015 inch)
Crankpin Standard Diameter . . . . . 63.47-63.49 mm
                            (2.4988-2.4998 inches)
Crankpin Fillet Radii . . . . 3.68-3.96 mm
                            (0.145-0.156 inch)
Rod Bearing Clearance
  on Crankpin . . . . 0.03-0.08 mm
                            (0.0012-0.0031 inch)

To reinstall crankshaft, reverse removal procedure. Notice that main bearing caps are numbered 1 through 7 with number 1 cap at front of engine. The caps are also marked with the serial number of the engine. Make certain thrust washers are installed with grooved side toward crankshaft. Tighten main bearing cap screws to 270 N·m (200 ft.-lbs.) torque. Refer to paragraph 88 for installation and sealing of the block bridge pieces.

CRANKSHAFT REAR OIL SEAL

Early 399 Models
With 6.3544 Engine

88. The lip type rear oil seal is contained in a one-piece seal retainer attached to the rear face of the engine block. The seal and retainer can be removed after splitting tractor between the engine and transmission as outlined in paragraph 152, then removing the clutch and flywheel. Unbolt and remove seal retainer, remove old seal from retainer, then clean retainer, engine block and crankshaft.

Inspect sealing surface of crankshaft for wear or damage. If flange is grooved, new seal may be pressed slightly further into retainer so that seal will contact an unworn area of crankshaft flange. Original location of seal is with seal approximately 2.2 mm (⅛ inch) from rear of retainer. Spring loaded lip of seal is toward inside of engine. If sealing surface of crankshaft is damaged, seal contacting surface may be polished or reground to minimum diameter of 133.17 mm (5.243 inches); however, **it is important that the last 4.8 mm (³⁄₁₆ inch) of crankshaft remain**

original diameter. The flywheel pilots on this surface of crankshaft.

When installing the bridge piece at either the front or rear, coat the metal to metal contact faces, especially the upper corners of block, with non-hardening sealer such as "Powerpart Hylosil." Insert new seal pieces at ends and inject sealer into grooves at ends of the bridge piece. Use a straightedge to make sure that front surface of the front bridge and the rear surface of the rear bridge are flush with surfaces of block.

Be sure that the two dowels used to locate the seal retainer are in position in block, coat gasket with sealer and position gasket on dowels. Lubricate seal and crankshaft, then carefully position seal and retainer over crankshaft, onto locating dowels against gasket and engine block. A guide (Massey-Ferguson No. PD.145) is available to expand seal when installing over crankshaft.

FLYWHEEL

Early 399 Models
With 6.3544 Engine

89. To remove the flywheel, first separate engine from transmission housing as outlined in paragraph 152 and remove the clutch. The flywheel is secured to the crankshaft flange by six evenly spaced cap screws.

> **CAUTION: Flywheel is only lightly piloted to crankshaft. Use caution when unbolting to prevent flywheel from falling and possibly causing injury.**

The starter ring gear can be renewed after the flywheel is removed. Heat new ring gear to 245° C (475° F) before installing in position. Do not overheat ring gear and heat gear evenly. Heat treatment of gear could be disturbed by improper or excessive heating. Install gear on flywheel with beveled teeth toward front. Allow gear to air cool. Refer to Fig. 146 and paragraph 154 or paragraph 156 if clutch surface of flywheel needs resurfacing.

It is important to properly time the flywheel to engine during installation. Turn crankshaft until the front piston is at Top Dead Center, then install flywheel so that the "TDC1" mark is visible in the timing inspection hole located in left side of flywheel housing. Tighten flywheel retaining cap screws to 100 N·m (74 ft.-lbs.) torque. Check flywheel runout with a dial indicator after flywheel is installed. Runout measurement at machined outer surface should not exceed 0.30 mm (0.012 inch). Runout measured at clutch surface must not exceed 0.025 mm (0.001 inch) for each 25 mm (1 inch) from flywheel center line to the point of dial indicator measurement. If runout is excessive, remove flywheel and check for burrs or foreign material.

OIL PAN

Early 399 Models
With 6.3544 Engine

90. The cast iron oil pan serves as part of the tractor frame and as attaching point for the tractor front support. To remove the oil pan, first drain engine oil and refer to paragraph 67 to split the front end away from the engine. Support oil pan, remove the attaching screws, then lower oil pan away from the engine.

Reinstall oil pan by reversing removal procedures. Tighten retaining screws to 20 N•m (15 ft.-lbs.) torque. Refer to paragraph 5 for installation, shimming and tightening front support screws.

OIL PUMP

Early 399 Models
With 6.3544 Engine

91. The rotor type oil pump is mounted on the lower surface of the engine block and is driven by the fuel injection pump drive gear (24—Fig. 90). The oil pump can be unbolted and removed after removing the oil pan as outlined in paragraph 90. Service parts for the oil pump are not available individually; however, pump can be disassembled for inspection. Clearance between the inner rotor and outer rotor should be 0.08-0.13 mm (0.003-0.005 inch) and outer rotor should have 0.15-0.33 mm (0.006-0.013 inch) clearance in pump body bore. End clearance of rotors in pump body should be 0.01-0.06 mm (0.0005-0.0025 inch). If pump is worn excessively or damaged, complete pump should be renewed.

RELIEF VALVE

Early 399 Models
With 6.3544 Engine

92. The oil pressure relief valve (2—Fig. 93) is contained in valve body housing (1). The relief valve housing can be unbolted and removed from lower surface of block after removing the oil pan as outlined in paragraph 90. The relief valve should open at 345-413 kPa (50-60 psi). Free length of the spring should be 58.74 mm (2.3125 inches) and spring should exert 1.58 kg (3.5 lbs.) when compressed to the installed length of 55.56 mm (2.1875 inches). Plunger (2) standard diameter is 18.18-18.21 mm (0.7158-0.7170 inch) and plunger should have 0.03-0.09 mm

(0.0010-0.0034 inch) clearance in bore. Parts of relief valve except body are available separately.

COOLING SYSTEM

Early 399 Models
With 6.3544 Engine

93. RADIATOR. Remove the precleaner, hood and both lower panels. Disconnect oil cooler supply and return tubes from models so equipped. Drain coolant and disconnect upper and lower radiator hoses from all models. Unbolt fan shroud from radiator and slide shroud toward rear over the fan and coolant pump. Remove the four screws attaching the radiator, then lift the radiator up from between fuel tank and fan shroud.

To reinstall, reverse removal procedure. Coolant capacity is approximately 20.1 L (21.2 qts.) for models without cab. Radiator cap is 0.75 bar (10 psi) rating.

94. THERMOSTAT. Two thermostats (2—Fig. 94) are contained under the coolant outlet (1). Thermostats can be removed after draining coolant to below level of thermostats, detaching upper radiator hoses, then unbolting thermostat housing. Each of the thermostats should begin to open at approximately 77-85° C (170-185° F) and be fully open at 92-98° C (198-208° F).

95. COOLANT (WATER) PUMP. To remove the coolant pump, first drain the cooling system, then remove the radiator as outlined in paragraph 93. Remove the radiator shroud, loosen drive belt, then unbolt and remove the fan and belt. Disconnect hoses from pump, remove pump mounting stud nuts and screws, then remove pump. Note that some of the retaining stud nuts are behind pulley.

To disassemble pump, first remove nut (4—Fig. 94) and pulley (5). Remove internal snap ring (6), then press shaft (10) and bearings (7) toward front, out of

Fig. 93—Exploded view of oil pump pressure relief valve.

1. Valve body
2. Plunger
3. Spring
4. Seat

Fig. 94—Exploded view of coolant pump and associated parts used on 6.3544 engine used in early 399 tractors.

1. Water outlet
2. Thermostats
3. Pump housing
4. Nut
5. Pulley
6. Snap ring
7. Bearings
8. Spacer
9. Seal
10. Shaft
11. Seal
12. Collar
13. Impeller
14. Mounting plate
15. Water connection
16. By-pass hose

Fig. 95—Press impeller onto shaft until clearance (D) is 0.69-0.89 mm (0.027—0.035 inch).

impeller (13) and housing (3). Remainder of disassembly will be evident.

To reassemble, reverse disassembly, observe the following. Press bearings (7) and spacer (8) onto shaft (10) with shielded side of bearings away from each other. Pack bearings, then half fill the space between bearings with high temperature grease. Press the complete bearing/shaft assembly into front of housing

until it bottoms. Support pulley end of pump and press impeller (13) onto shaft until distance (D—Fig. 95) is 0.69-0.89 mm (0.027-0.035 inch). Make certain that shaft rotates smoothly.

Reinstall coolant pump by reversing the removal procedure. Adjust fan belt tension to provide 10 mm (⅜ inch) deflection midway between pulleys.

SIX-CYLINDER 1006.6 ENGINE AND COMPONENTS

Six-cylinder (1006.6) engines are used in late 399 models. Refer to paragraph 35 and following for service to the three-cylinder engines installed in 340, 350, 355 and 360 models. Refer to paragraph 67 and following for service to the six-cylinder (6.3544) engines used in early 399 models. Refer to Condensed Service Data for original application. The engine serial number consists of up to 13 letters and numbers located as shown in Fig. 96. The numbers and letters constituting the engine serial number can be used to identify the engine family, country of manufacture and the year of manufacture. Refer to Fig. 97 for identifying code.

Fig. 96—View of the 6 Liter (365 cubic inch) displacement, six-cylinder engine used on late 399 models showing location of the engine serial number.

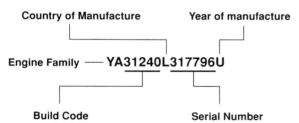

Fig. 97—The first two letters of the engine serial number identifies engine family, the second group is the build code. The letter approximately in the center identifies the country of manufacture and is followed by the engine serial number. The last letter indicates the year of manufacture. The illustration shown above identifies a 1006.6 engine (naturally aspirated, six-cylinder engine with 6 Liters/365 cubic inch displacement), built in Italy in 1990. The build code is 31240 and the serial number is 317796.

Engine family –
  YA = 1006.6 (6 cyl., 6 Liters)
Country of manufacture –
  A = Argentina
  B = Brazil
  C = Australia
  D = Germany
  E = Spain
  F = France
  G = Greece
  J = Japan
  K = Korea
  L = Italy
  M = Mexico
  N = U.S.A.
  P = Poland
  S = India
  T = Turkey
  U = United Kingdom
  W = Iran
  X = Peru
  Y = Yugoslavia
Year of manufacture –
  T = 1989
  U = 1990
  V = 1991
  W = 1992

FRONT END/ENGINE SPLIT

Late 399 Models With 1006.6 Engine

96. Block both rear wheels to prevent rolling, remove any front mounted equipment and any front weights. Remove hood, grille and side panels. Drain coolant from radiator and engine block. Remove the air cleaner and exhaust. Disconnect battery ground and front wiring harness to the headlights, horn and fuel tank. If necessary, drain fuel tank, then disconnect fuel feed and return lines. Disconnect top and bottom radiator hoses, hydraulic pipes to and from the oil cooler and hoses to the steering cylinders. On four-wheel-drive tractors, refer to paragraph 9 and remove drive shaft and shield. On all models, wedge between the front support and axle to prevent tipping and support the front axle, radiator and nose from above using a movable hoist. Support rear of tractor under transmission housing, then carefully remove stud nuts and screws attaching front support to the engine. Carefully roll the front axle, radiator and front assembly forward, away from engine.

Reassemble by reversing the splitting procedure. Refer to paragraph 5 for attaching front support to engine and selecting proper shims. Refer to paragraph 9 for installing drive shaft of four-wheel-drive models.

R & R ENGINE

Late 399 Models With 1006.6 Engine

97. To remove the engine, first refer to paragraph 96 and split tractor between front support and front of engine. Disconnect wires from engine components and position wires out of the way. Disconnect throttle and stop cables from the fuel injection pump, detach cables from support brackets and move the cables out of the way. Disconnect hydraulic lines between rear of tractor and front that would interfere with the removal of engine and cover all openings. On models

Fig. 98—On 1006.6 engines, tighten cylinder head retaining screws in the order shown. To prevent damage loosen fasteners in reverse of the or der shown. Long screws are used at locations (24 and 29). Studs should be installed temporarily in locations (15 and 20 or 25 and 30) when installing gasket and cylinder head to aid alignment and prevent sliding.

with cab, disconnect heater and air conditioning lines if so equipped. Remove gear lever trim plate and boot from models with cab. On all models, attach hoist to the engine, then remove the screws and nuts attaching the engine to transmission housing. Carefully move the engine forward away from the transmission.

To reinstall engine, reverse the removal procedure and observe the following. Install a M12 alignment stud approximately 100 mm (4 inches) long in each side of the transmission housing to assist in aligning engine to transmission. Turn flywheel to align clutch plate splines with transmission and pto input shaft splines. Install retaining screws and nuts after engine and transmission housing flanges are completely together. Tighten retaining screws and nuts to 115 N·m (85 ft.-lbs.) torque. Reverse removal procedure and refer to paragraph 5 for attaching front support to engine and selecting proper shims. Check and adjust clutch linkage as outlined in paragraph 151. Refer to paragraph 9 for installing drive shaft of four-wheel-drive models.

CYLINDER HEAD

Late 399 Models
With 1006.6 Engine

98. REMOVE AND REINSTALL. To remove cylinder head, first drain cooling system, remove hood and disconnect battery ground cable. Disconnect exhaust pipe and inlet pipe from manifolds. Remove the fuel pressure and return lines. Disconnect breather pipe and top radiator hose. Remove rocker arm cover and the rocker arm assembly as outlined in paragraph 103, then remove the push rods. Unbolt and remove inlet and exhaust manifolds from cylinder head. Unbolt and remove fuel injector assemblies from cylinder head. Remove the coolant by-pass hose. Loosen screws retaining cylinder head in reverse of sequence shown in Fig. 98. Unbolt and lift cylinder head. **Do not use sharp or hard tools between cylinder block and head and lay removed head on clean flat surface to prevent damage.**

Refer to specific paragraphs for servicing rocker arms, valves, seats and guides. Clean gasket surface of cylinder head and check for distortion or other

damage. Check flatness of cylinder head gasket surface using a straightedge and feeler gauge. Cylinder head should be renewed or resurfaced if surface is warped more than 0.13 mm (0.005 inch) across head or 0.25 mm (0.010 inch) from front to rear. Standard depth of cylinder head is 102.79-103.59 mm (4.047-4.078 inches). Cylinder head should **NOT** be resurfaced to less than 102.48 mm (4.035 inches) depth.

Guide studs should be temporarily installed in locations (25 and 30—Fig. 98) to align cylinder head and gasket and to prevent sliding while screws are installed. The cylinder head gasket should be marked "FRONT TOP" as shown in Fig. 99 and should be installed dry. Screws must be new or in very good condition. Check all screws for distortion and elongation using a straightedge along side of the threads. Discard any screw if any part appears to have a reduction in diameter. Lubricate the threads of screws, then install and tighten screws gradually and evenly to 110 N·m (81 ft.-lbs.) torque using the sequence shown (1 through 32—Fig. 98). Check to be sure that all are tightened to exactly 110 N·m (81 ft.-lbs.) torque, then tighten each fitting exactly the number of degrees specified for its length as follows. Short screws (5, 6, 11, 12, 16, 17, 21, 22, 26, 27, 31 and 32) should be turned 150° (2½ flats). Medium length screws (1, 2, 7, 8, 9, 10, 13, 14, 15, 18, 19, 20, 23, 25 and 28 and 30) should be turned 180° (3 flats or

Fig. 99—The cylinder head gasket should be installed with the words "FRONT TOP" up as shown. Guide studs (30) should be installed temporarily at screw positions (25 and 30—Fig. 98).

½ turn). The two longest screws (24 and 29) should be turned 210° (3½ flats).

Refer to Fig. 98 and the following for the specified amount of turning after all screws have been tightened to 110 N·m (81 ft.-lbs.) torque. Screws are listed in the order screws are to be tightened:

Screws 1 - 4 . . . . 180° (3 flats)
Screws 5 and 6 . . . . 150° (2½ flats)
Screws 7 - 10 . . . . 180° (3 flats)
Screws 11 and 12 . . . . 150° (2½ flats)
Screws 13 - 15 . . . . 180° (3 flats)
Screws 16 and 17 . . . . 150° (2½ flats)
Screws 18 - 20 . . . . 180° (3 flats)
Screws 21 and 22 . . . . 150° (2½ flats)
Screw 23 . . . . 180° (3 flats)
Screw 24 . . . . 210° (3½ flats)
Screw 25 . . . . 180° (3 flats)
Screws 26 and 27 . . . . 150° (2½ flats)
Screw 28 . . . . 180° (3 flats)
Screw 29 . . . . 210° (3½ flats)
Screw 30 . . . . 180° (3 flats)
Screws 31 and 32 . . . . 150° (2½ flats)

If screws are in good condition and if tightening is carefully and accurately done, the fasteners will retain the correct torque and will not require retightening. Special tools are available to assist in turning the screws the specified number of degrees, but marking the fastener and the cylinder head before turning will also be helpful. Tighten cap nuts retaining the rocker arm cover to 20 N·m (15 ft.-lbs.) torque. Screws retaining injection nozzles should be tightened to 12 N·m (9 ft.-lbs.) torque.

Refer to paragraph 103 when installing rocker arm assembly. Check and adjust valve clearance as outlined in paragraph 104. Install exhaust manifold without sealer. Inlet manifold gaskets should be installed with notch toward the top and straight edge toward the center. Remainder of reassembly procedure will be reverse of removal procedure.

VALVES AND SEATS

Late 399 Models
With 1006.6 Engine

99. When disassembling the cylinder head, keep valves in order so that they can be reinstalled in original location, if reused. Valve heads are marked with cylinder numbers consecutively from front to rear on some engines. Any replacement valves should be similarly marked before installation.

Inlet and exhaust valves originally seat directly in cylinder head; however, cylinder head may be machined and fitted with replacement seats. Inlet and exhaust valve face angles are 45° and seat angles are 46°. Inlet valves should be recessed 1.27-1.60 mm (0.050-0.063 inch) below surface of cylinder head and

should not be recessed more than 1.85 mm (0.073 inch). Exhaust valves should be recessed 1.28-1.83 mm (0.050-0.072 inch) below surface of cylinder head and should not be recessed more than 2.08 mm (0.082 inch). Clearance between valve stem and guide should not exceed 0.13 mm (0.005 inch) for inlet valves; 0.15 mm (0.006 inch) for exhaust valves. Standard valve stem diameter is 9.46-9.49 mm (0.3725-0.3735 inch) for inlet valves; 9.43-9.46 mm (0.371-0.372 inch) for exhaust valves. Refer to paragraph 100 for service to valve guides. Cup type valve stem seals are used on all valves.

VALVE GUIDES

Late 399 Models
With 1006.6 Engine

100. Inlet and exhaust valves operate in machined bores of valve guides that are pressed into the cylinder head. Clearance between valve stem and guide should not exceed 0.13 mm (0.005 inch) for inlet valves; 0.15 mm (0.006 inch) for exhaust valves. Standard valve guide inside diameter is 9.51-9.56 mm (0.3743-0.3764 inch). Length of inlet guide is 57.94 mm (2.281 inches); 61.10 mm (2.406 inches) for exhaust guides.

New service valve guides can be installed if original guides are worn. Press guides into cylinder head until top of guide is 15.10 mm (0.594 inch) above spring seating face. Be sure to reface valve seats after renewing valve guides. Cup type valve stem seals are used on all valves.

VALVE SPRINGS

Late 399 Models
With 1006.6 Engine

101. Valve springs, retainers and locks are interchangeable for inlet and exhaust valves. Cup type valve stem seals are used on all valves. Close wound damper coils should be installed toward cylinder head. Renew springs if discolored, distorted or if they fail to meet the following specifications. Single spring should exert 31.8-35.1 kg (70.1-77.3 lbs.) when compressed to 40.0 mm (1.57 inches).

CAM FOLLOWERS

Late 399 Models
With 1006.6 Engine

102. The mushroom type cam followers operate directly in the machined bores in cylinder block. Cam followers can be removed through bottom of cylinder

Fig. 100—Rocker arms must be assembled correctly to shaft as shown. Seal (1) is located in cylinder head recess to seal the oil passage to rocker arms. Rocker shaft brackets may be either aluminum or cast iron. Be sure of the type before tightening retaining screws. Screw (3) must engage hole in shaft.

block after removing timing gear housing as outlined in paragraph 112, the engine oil pan and camshaft as outlined in paragraph 113. Identify cam followers so they can be reinstalled in their original locations if reused.

The 18.99-19.01 mm (0.7475-0.7485 inch) diameter cam followers are available only in standard size and should have 0.04-0.09 mm (0.0015-0.00375 inch) diametral clearance in block bores. Install new cam followers if scored, pitted or excessively worn.

ROCKER ARMS

Late 399 Models
With 1006.6 Engine

103. The rocker arms and shaft assembly can be removed after removing the hood and rocker arm cover. Be careful not to damage the seal between the rocker arm cover and the inlet manifold or the seals around the cover retaining screws. The rocker arms are right and left hand units and should be assembled as shown in Fig. 100.

Rocker shaft diameter is 19.01-19.04 mm (0.7485-0.7495 inch) and the desired diametral clearance between shaft and rocker arms is 0.03-0.09 mm (0.0010-0.0035 inch). Renew shaft and/or rocker arms if clearance exceeds 0.13 mm (0.005 inch) or if either surface of shaft or bushing in rocker arm is rough.

Oil is supplied to the rocker arm shaft through the oil feed connection located between cylinders two and three on the rocker arm shaft. The oil feed connection is zinc and washers (2) are installed between oil feed connection and springs and between end rocker arms and retaining ring at ends. Make sure that seal ring (1) for base of oil feed connection is in place when installing rocker arm assembly. Rocker arm brackets must be installed with correct side toward top as shown in Fig. 100. Tighten screws retaining aluminum rocker shaft support brackets to 40 N·m (30 ft.-lbs.) torque. Tighten screws retaining cast iron rocker shaft support brackets to 75 N·m (55 ft.-lbs.) torque. Tighten rocker arm shaft retaining screws evenly, beginning in the center. Screws retaining the rocker arm cover are sealed with grommets in the cover. Tighten rocker cover retaining screws evenly to 20 N·m (15 ft.-lbs.) torque.

VALVE CLEARANCE

Late 399 Models
With 1006.6 Engine

104. Recommended valve clearance is 0.2 mm (0.008 inch) cold for inlet valves; 0.45 mm (0.018 inch) cold for exhaust valves. Static setting of all valves can be made from two crankshaft positions using the following procedure:

Remove the timing window plug from the left side of flywheel adapter housing and turn the flywheel until "TDC 1" mark on flywheel is aligned in center of hole. Check valves for front cylinder. Both valves should be closed if front cylinder is on compression stroke. If front two valves are tight (open), turn crankshaft exactly one complete revolution to set front piston on compression stroke.

If not equipped with hole in flywheel adapter or mark on flywheel, turn flywheel in normal direction of rotation until the exhaust valve for number 6 (rear) cylinder is just about closed and the inlet valve is just beginning to open. If not equipped with marks, mark crankshaft pulley so that crankshaft can be rotated exactly one complete revolution from this position after adjusting the first set of valves indicated in Fig. 101.

With front cylinder set at TDC on compression stroke, adjust the six valves indicated in Fig. 101. Turn the crankshaft exactly one complete revolution and adjust the six valves indicated in Fig. 102.

When finished with adjustment, install rocker arm cover. Screws retaining the cover are sealed with

Fig. 101—With TDC timing marks aligned and number 1 (front) piston on compression stroke, adjust the valves indicated. After adjusting these valves, turn crankshaft exactly one revolution (until marks again align) then adjust valves shown in Fig. 102.

Fig. 102—With TDC timing marks aligned and number 1 (front) piston on exhaust stroke, adjust the valves indicated.

grommets and should be tightened evenly to 20 N·m (15 ft.-lbs.) torque.

CRANKSHAFT PULLEY

Late 399 Models
With 1006.6 Engine

105. REMOVE AND REINSTALL. To remove the crankshaft pulley, first refer to paragraph 96 and split tractor between front support and front of engine. Loosen alternator and air conditioning compressor (if so equipped), then remove drive belts. Remove the three screws retaining thrust block (1—Fig. 103) to crankshaft, then withdraw thrust block and "O" ring (2). Pulley may be free to remove from crankshaft; however, it may be necessary to bump hub of pulley to separate the tapered locking rings (3 and 4). **Do not use a puller.** Using a puller would cause tapered rings to lock and crankshaft or pulley may be damaged. Remove tapered rings (3 and 4), spacer (5), pulley and key (6).

Install the crankshaft pulley, spacer (5), tapered rings (3 and 4), "O" ring (2) and thrust block (1).

Fig. 103—Crankshaft pulley and related parts. Refer to text for recommended removal procedure.

1. Thrust block
2. "O" ring
3. Outer taper ring
4. Inner taper ring
5. Spacer
6. Woodruff key

Install ring (4), which is cut away on outside before installing ring (3). Install and tighten the three retaining screws to 115 N·m (85 ft.-lbs.) torque.

TIMING GEAR COVER AND FRONT OIL SEAL

Late 399 Models
With 1006.6 Engine

106. To remove the timing gear cover, refer to paragraph 96 and remove the front axle, support and associated parts. Remove accessory drive belts, alternator and fan blades. Remove the crankshaft pulley as outlined in paragraph 105 and coolant pump as outlined in paragraph 126. Remove screws attaching the timing gear cover to the front of engine. Be careful to make sure that all of the screws are removed.

Press (do not pry) old crankshaft oil seal from cover bore, then press new seal into position with spring loaded lip toward inside. The crankshaft oil seal is directional and stamped with an arrow to indicate the direction of rotation of the engine crankshaft. Press new seal into bore until front surface of seal case is 9.3 mm (0.366 inch) from front surface of timing gear cover.

To install timing gear cover, first install a new seal into cover, then position gasket on front of engine. Position cover and install retaining screws loosely. Center timing gear cover around the crankshaft using special centering tool (PD163A available from Massey-Ferguson) or equivalent, then tighten the retaining screws and stud nuts to 22 N·m (16 ft.-lbs.) torque. If special centralizing tool is not available, the crankshaft pulley can be used carefully, but damage to the soft lip of the seal is likely. Install the crankshaft pulley as outlined in paragraph 105.

TIMING THE GEARS

Late 399 Models
With 1006.6 Engine

107. Before attempting to remove any of the timing gears, first remove the rocker arm cover and the rocker arms to avoid possible damage to pistons or valve train. The valves can (and probably will) hit the pistons if either the camshaft or crankshaft is turned independently of the other. The timing marks shown in Fig. 104 align only once every 18 revolutions due to the odd number of teeth on the idler gear.

To remove the timing gears or align marks on gears, refer to paragraph 96 and remove the front axle, support and associated parts, then remove the timing gear cover as outlined in paragraph 106. Refer to paragraphs 108, 109, 110 and 111 for service information regarding each of the four gears. With rocker

Fig. 104—Drawing of timing gears for 1006.6 engine showing timing marks aligned. Inset shows injection pump gear aligned with tooth marked "6" aligned for six-cylinder engine.

1. Crankshaft gear
2. Idler gear
3. Camshaft gear
4. Injection pump gear
5. Steering tandem pump

arms removed, crankshaft gear, camshaft gear and injection pump gear can be rotated as necessary to align timing marks before reinstalling idler gear with all timing marks aligned.

TIMING GEARS

Late 399 Models
With 1006.6 Engine

108. IDLER GEAR. Backlash should be at least 0.08 mm (0.003 inch) between the idler and any other gear. Gear should have 0.10-0.20 mm (0.004-0.008 inch) end play on shaft. The 63 tooth idler gear is retained on idler shaft by three screws and a washer.

Fig. 105—Idler gear hub should be installed with lubricating hole (H) up as shown.

Idler shaft diameter should be 50.70-50.74 mm (1.9960-1.9975 inches). Gear to shaft diametral clearance should be 0.058-0.119 mm (0.0023-0.0047 inch). The flanged bushings in idler gear must be reamed after installation to inside diameter of 50.80-50.82 mm (1.9998-2.0007 inches). Lubrication hole (H—Fig. 105) must be toward top as shown.

Make sure that timing marks on all gears are aligned as shown in Fig. 104 when idler gear is installed. Tighten idler gear retaining screws to 44 N·m (33 ft.-lbs.) torque.

109. CAMSHAFT GEAR. Gears are keyed to camshaft and crankshaft and valve timing will be correct if timing marks are properly aligned as shown in Fig. 104. There is no adjustment provided and if valve timing is incorrect, gears are one or more teeth out of correct mesh. The valves can (and probably will) hit the pistons if either the camshaft or crankshaft is turned independently of the other, unless the rocker arms are removed.

Marked teeth of camshaft gear (3) and the idler gear (2) can be viewed after removing the timing gear cover as outlined in paragraph 106. The camshaft gear can be pulled from camshaft after removing the center screw. The gear has two threaded holes for attaching a puller.

Because the idler gear has an odd number of teeth, timing marks will only occasionally align. Since marks do not align every other revolution, alignment is simplified by removing rocker arms and idler gear (2), moving gears (1, 3 and 4) as necessary to align marks, then reinstalling idler (2). The injection pump drive must also be timed as outlined in paragraph 110.

Screw retaining the camshaft gear (3) should be tightened to 78 N·m (58 ft.-lbs.) torque and screws retaining idler gear (2) should be tightened to 44 N·m (33 ft.-lbs.) torque.

110. INJECTION PUMP DRIVE GEAR. Gear is keyed to the fuel injection pump shaft and timing will be correct if the timing marks are properly aligned as shown in Fig. 104. There is no adjustment provided and if timing is incorrect, gears are one or more teeth out of correct mesh.

Marked teeth of timing gears (1, 2, 3 and 4) are visible after removing the timing gear cover as outlined in paragraph 106. The injection pump drive gear is marked for installation on both four and six-cylinder engines. Use the timing mark stamped "6" for this six-cylinder engine. The injection pump gear can be pulled from shaft after removing the center retaining nut and washer. The gear has two threaded holes for attaching a puller.

Because the idler gear has an odd number of teeth, timing marks will only occasionally align. Since marks do not align every other revolution, alignment is simplified by removing rocker arms and idler gear (2), moving gears (1, 3 and 4) as necessary to align marks, then reinstalling idler (2). The valves can (and probably will) hit the pistons if either the camshaft or crankshaft is turned independently of the other, unless the rocker arms are removed.

Nut retaining the injection pump gear (4) should be tightened to 80 N·m (59 ft.-lbs.) torque and screws retaining idler gear (2) should be tightened to 44 N·m (33 ft.-lbs.) torque.

111. CRANKSHAFT GEAR. The crankshaft gear and other timing gears are accessible to inspection, removal or service after removing the timing gear cover as outlined in paragraph 106 and the idler gear as outlined in paragraph 108. The crankshaft gear is keyed in place and fits the shaft with a transitional fit, 0.020 mm (0.0008 inch) tight to 0.025 mm (0.0010 inch) loose. If the old gear is loose on shaft, it may be possible to pry gear from shaft, but a puller may be required. If necessary to remove crankshaft gear with a puller, it is suggested to first remove the crankshaft.

TIMING GEAR HOUSING

Late 399 Models With 1006.6 Engine

112. REMOVE AND REINSTALL. To remove the timing gear housing, it is first necessary to remove the steering tandem pump as outlined in paragraph 29, the timing gear cover as outlined in paragraph 105 and the injection pump as outlined in paragraph 143. Remove the idler gear and camshaft gear as outlined in paragraphs 108 and 109. The housing can be removed without removing the oil pan by carefully cutting the gasket, then sealing the trimmed front section of oil pan gasket when reassembling. Removal, cleaning, sealing and reassembly may be easier if the oil pan is removed. If oil pan is removed, the idler gear hub can be reattached to front of block with the three gear retaining screws while the housing is removed. Remove screws attaching the timing gear housing to front of block and oil pan, then carefully remove housing. Be careful to make sure all screws are removed and do not drop idler gear hub or camshaft thrust washer.

If new housing is installed, mark rear of new housing with injection pump marks in exactly same location as old housing.

If oil pan is removed, temporarily attach idler gear hub to front of block with oil hole toward top using the three retaining screws. Make sure that camshaft thrust washer is properly located in block recess, install gasket between block and housing, then install timing gear housing to engine block.

If oil pan is not removed, trim new oil pan gasket to replace damaged section at front. The housing can be installed and successfully sealed to the oil pan without removing the oil pan, but extreme care must be exercised. Make sure that camshaft thrust washer is properly located in block recess and install gasket between block and housing. Locate idler gear hub with oil hole to top while positioning timing gear housing to front of engine and temporarily install the three screws to retain and align hub.

Install and tighten the four screws around idler gear hub to 22 N·m (16 ft.-lbs.) torque. Refer to paragraph 143 and install the fuel injection pump. Install camshaft gear and idler gear, aligning the timing marks as outlined in paragraphs 108 and 110. Refer to paragraph 105 and install the timing gear cover. Remainder of assembly is reverse of disassembly.

CAMSHAFT

Late 399 Models With 1006.6 Engine

113. To remove the camshaft, first remove the rocker arm assembly as outlined in paragraph 103, remove push rods and separate the front axle and support from the front of the engine as described in paragraph 96. Unbolt and remove the fuel lift pump from the right side of engine block. Remove engine oil pan, then refer to paragraph 112 and remove the timing gear housing. Push the cam followers up away from cam lobes and withdraw camshaft. Remove cam followers from below and thrust washer (Fig. 106) from front as camshaft is withdrawn from block. Identify cam followers for installation in the same location if they are to be reinstalled.

Cam lift is 7.627-7.69 mm (0.2999-0.3029 inch) for inlet lobes; 7.71-7.798 mm (0.3036-0.3066 inch) for exhaust lobes. Refer to paragraph 102 for service to cam followers. The camshaft is supported by four journals which operate directly in unbushed cylinder

Fig. 106—Dowel must engage hole in camshaft thrust washer and thrust washer must be seated fully in recess of block before installing the timing gear housing.

block bores. Desired clearance for all journals is 0.06-0.14 mm (0.0025-0.0055 inch). Refer to the following for standard journal diameters.

Journal # 1
    Diameter. . . . . 50.71-50.74 mm
                        (1.9965-1.9975 inches)

Journal # 2
    Diameter. . . . . 50.46-50.48 mm
                        (1.9865-1.9875 inches)

Journal # 3
    Diameter. . . . . 5.020-50.23 mm
                        (1.9765-1.9775 inches)

Journal # 4
    Diameter. . . . . 49.95-49.98 mm
                        (1.9665-1.9675 inches)

Camshaft end play is controlled by the thickness of the thrust washer (Fig. 106). Thrust washer should be 5.49-5.54 mm (0.216-0.218 inch) thick and should limit camshaft end play to 0.10-0.41 mm (0.004-0.016 inch). The thrust washer should fit in block recess no more than 0.05 mm (0.002 inch) below flush with front face of block, but should not protrude more than 0.08 mm (0.003 inch) above. When reassembling, be sure that thrust washer fits recess correctly and engages locating pin.

When reassembling, insert cam followers into bores of block from below and slide camshaft into block from front. Install camshaft thrust washer (Fig. 106) and timing gear housing as outlined in paragraph 112. Refer to paragraph 104 for adjusting valve clearance.

ROD AND PISTON UNITS

Late 399 Models With 1006.6 Engine

114. Connecting rod and piston units are removed from above after removing the cylinder head as outlined in paragraph 98, engine oil pan as outlined in paragraph 121, the oil strainer and suction pipe. Remove carbon and wear ridge from top of cylinder bore, unbolt and remove rod bearing cap, then push rod and piston assembly out of cylinder. Make sure that cylinder number is stamped connecting rod, cap and piston.

It is important that the top of piston is correct distance from the top face of cylinder block when assembled. Pistons are available with top surfaces different distances from piston pin and also "un-topped," which require machining the piston top to the correct height. Refer to paragraph 115.

Pistons should be assembled to connecting rods as shown in Fig. 107. Top of piston is marked with arrow or "FRONT" indicating correct installation. Make

Fig. 107—Notch (N) in connecting rod for tang of bearing insert should be on same side as lug (L) of piston boss as shown.

sure that notches (N) on rod and cap are together. Connecting rod nuts should be tightened to 125 N·m (92 ft.-lbs.) torque.

PISTONS, RINGS AND SLEEVES

Late 399 Models With 1006.6 Engine

115. The aluminum alloy, cam ground pistons are supplied in standard size only. The combustion chamber is offset in piston toward injection nozzle side of engine. Piston is marked with an arrow or "FRONT" indicating front of engine to facilitate assembly. Additionally, the combustion chamber is offset toward injection pump side of engine.

Assemble piston to connecting rod with notch (N—Fig. 107) for tang of connecting rod bearing insert on same side as lug (L) on underside of piston pin boss.

It is important that the top of piston is 0.14-0.36 mm (0.005-0.014 inch) above the top surface of the cylinder block when assembled. Pistons are available with top surfaces different distances from piston pin and also "un-topped," which require machining the piston top to the correct height. Install the connecting rod and piston assembly to the crankshaft and in cylinder block, but without piston rings, mark the top edge of cylinder block with piston at TDC, then remove and machine top of piston as required. After machining top of piston, top inner edge of the combustion chamber in piston should be rounded slightly and top of piston should be stamped with cylinder number and arrow or "FRONT" mark.

Three piston rings are used. See Fig. 107A. The first (top) ring is barrel faced and internally chamfered with molybdenum insert. The internal chamfer of the top ring, side of ring marked "T", "TOP" or with the manufacturer's mark should be installed toward top of piston. The second ring is not stepped or chamfered, but face of ring is tapered. The side of the second ring marked "T", "TOP" or with the manufacturer's mark should be toward top of piston. The bottom (third), chrome plated oil control ring is nondirectional with a coil spring expander. Install the expander in lower groove first and join ends with the locating wire. Install the ring over the expander, making certain that joint of expander is 180° from (opposite) gap of ring. Piston and ring specifications are as follows.

Three-Ring Pistons
Ring Width –
Top . . . . 2.48-2.49 mm
(0.097-0.098 inch)
Second . . . . 2.48-2.49 mm
(0.097-0.098 inch)
Bottom . . . . 3.98-3.99 mm
(0.1566-0.1571 inch)
Ring Side Clearance in Groove –
Top . . . . 0.08-0.11 mm
(0.003-0.004 inch)
Second . . . . 0.06-0.09 mm
(0.002-0.003 inch)
Bottom . . . . 0.04-0.08 mm
(0.002-0.003 inch)
Ring End Gap –
Top . . . . 0.40-0.85 mm
(0.016-0.033 inch)
Second . . . . 0.30-0.76 mm
(0.012-0.030 inch)
Bottom . . . . 0.38-0.84 mm
(0.015-0.033 inch)

Original cylinder bore diameter is 100.00-100.03 mm (3.937-3.938 inches), install new liner if taper, out-of-round or wear exceeds 0.25 mm (0.010 inch). Suitable sleeve removal and installing tools are required to renew sleeves. Thoroughly clean and inspect cylinder block bores and new sleeves before installing, because even the slightest burr or dirt can cause distortion of new sleeve as it is pressed into block. Measure height of liner (sleeve) flange and

Fig. 107A—Cross-sectional view of piston rings showing correct assembly position.

depth of block recess. Cylinder liner flange should be from 0.10 mm (0.004 inch) below surface of block to 0.10 (0.004 inch) above block surface. Shims are available which may be installed under flange if necessary to raise sleeve. Make sure that block recess is clean. Standard depth of block recess is 3.81-3.91 mm (0.150-0.154 inch).

Lubricate cylinder block bores with new engine oil just prior to installing new sleeves. Flanged, prefinished, service sleeves should fit block bores from 0.03 mm (0.001 inch) loose to 0.03 mm (0.001 inch) tight. Correct installed diameter of prefinished service liners is 100.04-100.06 mm (3.9385-3.9395 inches). Allow sufficient time for cylinder liner to stabilize, then check installed diameter of sleeves carefully at several locations from top to bottom of bore and in several different locations across bore.

Unfinished cylinder liners similar to those used in original production are available for service. Unfinished liners are 0.03-0.08 mm (0.001-0.003 inch) interference fit in cylinder block bores and require heavy pressing equipment. Accurate boring equipment should be used to machine installed sleeve to original cylinder bore diameter of 100.00-100.03 mm (3.937-3.938 inches). Sleeves with oversize outside diameter are also available for service. Standard block bore is 104.20-104.23 mm (4.103-4.104 inches).

PISTON PINS

Late 399 Models With 1006.6 Engine

116. The full floating piston pin is retained in piston bosses by snap rings.

Piston pin diameter is 34.920-34.925 mm (1.3748-1.3750 inches). The piston pin should have 0.003-0.014 mm (0.0001-0.0006 inch) clearance in piston and should have 0.020-0.043 mm (0.0008-0.0017 inch) clearance in bushing at top of connecting rod. To remove piston pin, remove snap rings from both ends, then heat the piston to 40-50° C (100-120° F). Pin should slide out after heating.

Bushing in top of rod is renewable. Be sure that oil hole in bushing is aligned with oil hole in top of rod bore. Bushing must be reamed to provide correct clearance for pin. Refer to paragraph 86 for installation of new pin bushing.

CONNECTING RODS AND BEARINGS

Late 399 Models With 1006.6 Engine

117. Connecting rod crankpin bearings are steel backed, aluminum/tin faced type renewable from below after removing the oil pan and rod bearing caps. The tangs of bearing inserts must engage notches (N—Fig. 107) of rod and bearing cap and must both be on the same side. If new connecting rod is installed, cylinder number should be stamped on rod and cap.

Bushing in top of rod for piston pin is renewable. Be sure that oil hole in bushing is aligned with oil hole in top of rod bore while pressing new bushing into position. Size new, installed bushing to 34.94-34.96 mm (1.3758-1.3765 inches). Clearance between piston pin and correctly sized bushing should be 0.020-0.043 mm (0.0008-0.0017 inch). Pin bushing must also be parallel to bore for crankshaft crankpin. Connecting rod length between centers is 219.05-219.10 mm (8.624-8.626 inches).

Crankshaft crankpin diameter is 63.47-63.49 mm (2.499-2.500 inches) new. Bearing insert are available in standard size and 0.25, 0.50 and 0.75 mm (0.010, 0.020 and 0.030 inch) undersizes. If crankpin is resized, crankpin fillet radii should be 3.68-3.96 mm (0.145-0.156 inch). Rod bearing diametral clearance on crankpin should be 0.025-0.076 mm (0.001-0.003 inch).

CRANKSHAFT AND BEARINGS

Late 399 Models
With 1006.6 Engine

118. The crankshaft is supported in seven steel backed, aluminum/tin or lead/bronze faced, renewable bearing inserts. Projections (tangs) on bearing insert must engage milled slots in block and main bearing cap. Tangs of inserts in block and cap must both be on same side of engine.

To remove the crankshaft, first drain coolant and engine oil, then remove engine as outlined in paragraph 97. Remove oil pan, then refer to paragraph 112 and remove timing gear housing. Remove clutch, flywheel, engine adapter plate and crankshaft rear oil seal. Remove connecting rod caps, main bearing caps, front and rear block bridge pieces, then lift crankshaft from cylinder block.

The upper and lower halves of main bearing inserts are not interchangeable. The upper insert halves are slotted to provide pressure lubrication to crankshaft and connecting rods. Inserts are interchangeable in pairs (upper and lower halves) for all journals except the center main bearing. Main bearing inserts are available in standard size and 0.25, 0.50 and 0.75 mm (0.010, 0.020 and 0.030 inch) undersizes. If crankshaft main journal is resized, journal fillet radii should be 3.68-3.96 mm (0.145-0.156 inch).

Crankshaft end play is controlled by renewable thrust washers at front and rear of center main bearing. The cap half is prevented from turning by the tab which fits in a machined notch of cap. Block half of washer can be rolled from position when cap is removed. Crankshaft end play should not exceed 0.51 mm (0.020 inch). Thrust washers are available in 0.18 mm (0.007 inch) oversize as well as standard

thickness. One set (top and bottom) of oversize thrust washers may be installed on one side (front or rear) in combination with a standard size set on other side to provide oversize adjustment of 0.18 mm (0.007 inch). Two sets of oversize thrust washers may be used to provide adjustment of 0.36 mm (0.014 inch).

Check crankshaft against the values which follow:

Main Journal Standard Diameter . 76.16-76.18 mm
(2.998-2.999 inches)
Main Journal Fillet Radii . . . . 3.68-3.96 mm
(0.145-0.156 inch)
Main Bearing Clearance . . . . 0.046-0.107 mm
(0.0018-0.0042 inch)
Crankshaft End Play. . . . . 0.05-0.38 mm
(0.002-0.015 inch)
Crankpin Standard Diameter. . . . . 63.47-63.49 mm
(2.499-2.500 inches)
Crankpin Fillet Radii . . . . 3.68-3.96 mm
(0.145-0.156 inch)
Rod Bearing Clearance
on Crankpin . . . . 0.025-0.076 mm
(0.001-0.003 inch)

To reinstall crankshaft, reverse removal procedure. Notice that main bearing caps are numbered 1 through 7 with number 1 cap at front of engine. The caps are also marked with the serial number of the engine. Make certain thrust washers are installed with grooved side toward crankshaft. Tighten main bearing cap screws to 270 N·m (200 ft.-lbs.) torque. Refer to paragraph 119 for installation and sealing of the block bridge pieces.

CRANKSHAFT REAR OIL SEAL

Late 399 Models
With 1006.6 Engine

119. The lip type rear oil seal is contained in a one-piece seal retainer attached to the rear face of the engine block. The seal and retainer can be removed after splitting tractor between the engine and transmission as outlined in paragraph 152, then removing the clutch and flywheel. Unbolt and remove seal retainer, remove old seal from retainer, then clean retainer, engine block and crankshaft.

Inspect sealing surface of crankshaft for wear or damage. If flange is grooved, new seal may be pressed slightly further into retainer so that seal will contact an unworn area of crankshaft flange. Original location of seal is with seal approximately 2.2 mm (0.09 inch) from rear of retainer. Spring loaded lip of seal is toward inside of engine. If sealing surface of crankshaft is damaged, seal contacting surface may be polished or reground to minimum diameter of 133.17 mm (5.243 inches); however, **it is important that the last 4.8 mm ($^3/_{16}$ inch) of crankshaft remain original diameter.** The flywheel pilots on this surface of crankshaft.

When installing the bridge piece at either the front or rear, coat the metal to metal contact faces, especially the upper corners of block, with non-hardening sealer such as "Powerpart Hylosil." Insert new seal pieces at ends and inject sealer into grooves at ends of the bridge piece. Use a straightedge to make sure that front surface of the front bridge and the rear surface of the rear bridge are flush with surfaces of block. Tighten bridge piece to cylinder block screws to 9 N·m (72 in.-lbs.) torque. Tighten rear oil seal housing to engine block screws to 22 N·m (16 ft.-lbs.) torque and screws retaining rear oil seal housing to bridge piece to 18 N·m (13 ft.-lbs.) torque.

Be sure that the two dowels used to locate the seal retainer are in position in block, coat gasket with sealer and position gasket on dowels. Lubricate seal and crankshaft, then carefully position seal and retainer over crankshaft, onto locating dowels against gasket and engine block. A guide (Massey-Ferguson No. PD.145D) is available to expand seal when installing over crankshaft.

FLYWHEEL

Late 399 Models
With 1006.6 Engine

120. To remove the flywheel, first separate engine from transmission housing as outlined in paragraph 152 and remove the clutch. The flywheel is secured to the crankshaft flange by six evenly spaced cap screws.

> **CAUTION: Flywheel is only lightly piloted to crankshaft. Use caution when unbolting to prevent flywheel from falling and possibly causing injury.**

The starter ring gear can be renewed after the flywheel is removed. Heat new ring gear to 245° C (475° F) before installing in position. Do not overheat ring gear and heat gear evenly. Heat treatment of gear could be disturbed by improper or excessive heating. Install gear on flywheel with beveled teeth toward front. Allow gear to air cool. Refer to Fig. 146 and paragraph 154 or paragraph 156 if clutch surface of flywheel needs resurfacing.

It is important to properly time the flywheel to engine during installation. Turn crankshaft until the front piston is at Top Dead Center, then install flywheel so that the "TDC1" mark is visible in the timing inspection hole located in left side of flywheel housing. Tighten flywheel retaining cap screws to 105 N·m (77 ft.-lbs.) torque. Check flywheel runout with a dial indicator after flywheel is installed. Runout measurement at machined outer surface should not exceed 0.30 mm (0.012 inch). Runout measured at clutch surface must not exceed 0.025 mm (0.001 inch) for each 25 mm (1 inch) from flywheel center line to the point of dial indicator measurement. If runout is excessive, remove flywheel and check for burrs or foreign material.

OIL PAN

Late 399 Models
With 1006.6 Engine

121. The cast iron oil pan serves as part of the tractor frame and as attaching point for the tractor front support. To remove the oil pan, first drain engine oil and refer to paragraph 96 to split the front end away from the engine. Support oil pan, remove the attaching screws, then lower oil pan away from the engine.

Reinstall oil pan by reversing removal procedures. Tighten retaining screws to 22 N·m (16 ft.-lbs.) torque. Refer to paragraph 5 for installation, shimming and tightening front support screws.

OIL PUMP

Late 399 Models
With 1006.6 Engine

122. The rotary type oil pump is mounted on front main bearing cap and is driven from the crankshaft timing gear through an idler. To remove the pump, first remove the oil pan as outlined in paragraph 121. Remove the oil delivery (inlet) pipe and the oil suction pipe. Unbolt and remove the front main bearing cap with the oil pump attached. Be careful not to damage the main bearing inserts. The main bearing upper insert may release and fall later if not secured. Pump can be unbolted from main bearing cap after removing the idler gear.

The pump idler gear is retained on shaft by a snap ring. Cover can be removed from pump and idler gear can be removed, but drive gear and inner rotor should not be removed from shaft. Clearances can be measured without further disassembly. Service parts for the oil pump are not available individually; however, pump can be disassembled for inspection. Clearance between the inner rotor and outer rotor should be 0.04-0.13 mm (0.0015-0.0050 inch) and outer rotor should have 0.15-0.34 mm (0.006-0.013 inch) clearance in pump body bore. End clearance of inner rotor in pump body should be 0.05-0.12 mm (0.002-0.005 inch). End clearance of outer rotor in pump body should be 0.04-0.11 mm (0.0015-0.0044 inch). If pump is worn excessively or damaged, complete pump should be renewed. Tighten oil pump cover screws to 28 N·m (21 ft.-lbs.) torque.

Reinstall pump by reversing removal procedure. Tighten oil pump to main bearing cap screws to 22

N·m (16 ft.-lbs.) torque and main bearing retaining screws to 265 N·m (196 ft.-lbs.) torque.

RELIEF VALVE

Late 399 Models
With 1006.6 Engine

123. The oil pressure relief valve (2—Fig. 108) is contained in valve body housing (1). The relief valve housing can be unbolted and removed from lower surface of block after removing the oil pan as outlined in paragraph 121. The relief valve should open at 345-413 kPa (50-60 psi). Relief valve spring should exert 1.3-1.9 kg (2.9-4.2 lbs.) when compressed to the installed length of 55.6 mm (2.2 inches). Plunger (2) standard diameter is 18.16-18.18 mm (0.715-0.716 inch) and plunger should have 0.06-0.11 mm (0.002-0.004 inch) clearance in bore. Parts of relief valve except body are available separately.

COOLING SYSTEM

Late 399 Models
With 1006.6 Engine

124. RADIATOR. Remove the precleaner, hood and both lower panels. Disconnect oil cooler supply and return tubes from models so equipped. Drain coolant and disconnect upper and lower radiator hoses from all models. Unbolt fan shroud from radiator and slide shroud toward rear over the fan. Remove the four screws attaching the radiator, then lift the radiator up from between fuel tank and fan shroud.

To reinstall, reverse removal procedure. Coolant capacity is approximately 23.0 L (24 qts.) for models without cab. Radiator cap is 0.75 bar (10 psi) rating.

125. THERMOSTAT. Two thermostats are contained under the coolant outlet. Thermostats can be removed after draining coolant to below level of thermostats, detaching upper radiator hoses, then unbolting thermostat housing. Each of the thermostats should begin to open at approximately 77-85° C (170-185° F) and be fully open at 92-98° C (198-208° F).

126. COOLANT (WATER) PUMP. The coolant pump is gear driven and sealed to the timing gear cover with two "O" rings (1 and 2—Fig. 109).

To remove the coolant pump, first drain the cooling system, then remove the radiator as outlined in paragraph 124. Remove the radiator shroud, then remove fan belts and fan. Disconnect coolant inlet hose from pump, then remove pump mounting screws and stud nuts. Withdraw pump forward out of timing gear cover.

To disassemble pump, remove the three studs (3—Fig. 110) and "O" ring (2), then unbolt and remove front cover (10). Use a suitable puller to remove the drive gear (4), then press shaft and bearing assembly (9), together with impeller (8), out toward front of housing (6). Support impeller and press shaft out of impeller hub.

Inspect all parts for wear or damage and renew as necessary. Support housing (6) and press shaft and bearing (9) into housing by pressing only on outer edge of bearing, not shaft. Front of bearing should be flush with counterbore in housing for coolant seal (7). **Do not lubricate or otherwise contaminate coolant seal (7).** Press coolant seal into housing, by applying force to the outer edge of seal. Hold seal in

Fig. 108—Exploded view of oil pump pressure relief valve.

1. Valve body	3. Spring
2. Plunger	4. Seat

Fig. 109—The water pump is driven by the timing gears. Make sure that both "O" rings (1 and 2) are removed and new "O" rings are used when installing pump.

Fig. 110—Cross section of gear driven
water pump used on 1006.6 engines installed in late 399 tractor models.

2. "O" ring
3. Studs
4. Drive gear
5. Oil seal
6. Pump housing
7. Water seal
8. Impeller
9. Shaft and bearing
10. Cover
11. Gasket

Fig. 111—Cross section of fan hub showing correct distance (F) which should be 110.3 mm (4.34 inches).

position for about 10 seconds to make sure that seal
does not move out of bore. Lubricate oil seal (5), then
press into housing with spring loaded lip toward rear
(gear). Support rear end of shaft (9), then press impeller (8) onto shaft until distance (A) between front
cover surface of housing and impeller hub is 8.1-8.5
mm (0.31-0.33 inch). Support shaft (9) and press gear
(4) onto shaft until distance (B) is 0.6-2.6 mm (0.02-
0.10 inch). Install pump mounting studs (3) in housing. Make certain that shaft rotates smoothly.

Reinstall coolant pump by reversing the removal
procedure.

127. FAN. The cooling fan is attached to a separate
fan drive hub shown in Fig. 111. Shaft should have
less than 0.25 mm (0.010 inch) end play. Distance (F)
should be 110.3 mm (4.34 inches) when pressing fan
hub on shaft. Tighten mounting screws to 44 N·m (33
ft.-lbs.) torque.

TURBOCHARGER

OPERATION

355 and 360 Models

128. The Perkins AT 3.1524 engines used in 355 and 360 tractor models are equipped with exhaust driven Holset HID turbochargers.

The turbocharger supplies air to the inlet manifold at above atmospheric pressure. The additional air entering the combustion chamber permits an increase in the amount of fuel that can be injected and burned, increasing the engine's power. Using the engine exhaust to power the compressor increases the engine's flexibility, enabling it to perform with the economy of a smaller engine on light loads yet permitting a substantial horsepower increase at full load. Horsepower loss because of altitude or atmospheric changes is also greatly reduced.

The turbocharger contains a rotating shaft with an exhaust turbine wheel on one end and a centrifugal air compressor on the other end. The exhaust turbine operates in the exhaust passage and is rotated by the exiting exhaust gasses. The compressor impeller is turned by the rotating exhaust turbine via a connecting shaft. The rotating assembly is precisely balanced and capable of rotating at very high speeds. Shaft bearings are full floating sleeve type and the unit is both lubricated and cooled by a flow of engine oil.

Do not operate the engine, even for a short time, without adequate (abundant) lubrication to the turbocharger unit. When a new or rebuilt turbocharger is first installed, the engine has not been run for a month or more, or if a new oil filter has been installed, shut fuel OFF, then crank engine with the starter until oil pressure is at a safe level. Turn fuel ON after oil pressure reaches normal, start engine, then operate engine at slow idle speed for at least two minutes before opening throttle or putting load on engine.

Do not operate engine at wide open throttle immediately after starting engine. Allow engine to idle for a while before stopping engine, to slow turbocharger speed before stopping engine.

If the engine stops suddenly while operating at heavy load and high speed, it is important to restart the engine immediately. Lack of lubrication, high rotational speed of the turbocharger and high temperature can quickly damage the turbocharger.

Maintain air filters and connections in good condition. Check system regularly and clean, renew or repair inlet air system as needed. The increased air flow of turbocharged engines may cause filters to become clogged more often than expected. Openings that permit entrance of unfiltered (dirty) air can quickly damage any engine, but damage will occur sooner on turbocharged engines because of the increased volume of air used.

Make sure exhaust pipe is covered and the air filter is connected when transporting. If the exhaust is equipped with a weathercap, tape the cap closed. If not, tape the pipe closed. If openings are not covered, air passing the open pipe may cause turbocharger to spin without adequate lubrication and damage bearings and other parts.

SERVICE

355 and 360 Models

129. REMOVE AND REINSTALL. To remove the turbocharger, first remove the tractor hood and muffler. Disconnect oil lines and air connections from turbocharger. Unbolt turbocharger from the exhaust manifold, then lift unit from tractor.

When installing, attach turbocharger to exhaust manifold using a new gasket. Attach the oil drain pipe to the turbocharger, pour 110-114 ml (4-5 fl.-oz.) of engine oil through the oil inlet port and turn turbocharger by hand to circulate the oil. Attach the oil inlet pipe, but do not tighten the screws. Operate the engine stop control while cranking the engine with starter until oil flows from the loosened inlet pipe connection, then tighten the oil supply connection. Attach air inlet connections, start engine and check for oil leaks. Complete assembly by reversing removal procedure.

130. INSPECTION AND OVERHAUL. Exchange turbocharger units are available for service, or a qualified technician can overhaul the unit if parts are available.

Refer to Fig. 112 for exploded view of Holset HID T31 turbocharger. Maximum boost pressure is approximately 0.5 bar (7.0 psi) at 2250 engine rpm for three-cylinder models.

Before disassembling any turbocharger, visually inspect turbine wheel and compressor impeller blades for damage by looking through housing and end openings. Check to see if the compressor impeller or turbine blades have rubbed on the housings. Check to see if housings are wet with oil or have carbon deposits, possibly indicating engine wear or damage. Check turbocharger castings for cracks. Check lubricating oil inlet and outlet ports for carbon build up or obvious damage to bearings. Rotate shaft by hand and check for freedom of rotation, without excessive noise or play.

Mark across compressor housing (1), bearing housing (13) and turbine housing (17) before disassembling, to aid alignment when reassembling. Clamp

Fig. 112—Exploded view of Holset HID turbocharger used on three-cylinder AT3.1524 engines.

1. Compressor housing
2. Locknut
3. Compressor impeller
4. Back plate
5. Seal ring
6. Clamp plates
7. Seal ring
8. Thrust collar
9. Thrust bearing
10. Bearing retainers
11. Bearings
12. Seal ring
13. Center housing
14. Lock plate
15. Shroud
16. Turbine shaft
    & wheel
17. Turbine housing
18. Seal ring
20. Insert

the turbocharger exhaust (mounting) flange in a vise and remove cap screws, lock plates and retaining plates, then remove the compressor housing (1). Remove cap screws, lock plates and retaining plates from other side, then lift turbocharger from the turbine housing (17).

> **CAUTION: Do not rest weight of any parts on impeller or turbine blades. Weight of only the turbocharger is enough to damage the blades.**

Use an approved holding fixture to hold exhaust turbine and shaft (16) while removing locknut (2) with a "T" handle wrench. **Turbine shaft may be easily bent if nut is removed without using "T" handle or double universal joint.** Lift compressor impeller (3) from shaft, then withdraw exhaust turbine and shaft (16) from center housing (13) and bearings (11). Shroud (15) is not retained and will fall free when shaft is removed. Unbolt and remove back plate (4) and outside snap rings (10), then carefully withdraw bearings (11). Be careful not to damage bearings (11) or surface of center housing (13) when removing retaining snap rings (10) or bearings. The two center snap rings do not need to be removed unless damaged or unseated. Always install new snap rings if removed.

Clean all parts in cleaning solution that is not harmful to aluminum. A bristle (not wire) brush and plastic or wood scraper can be used to remove sof-

tened deposits. Use extreme caution, when cleaning, to prevent parts from being nicked, scratched or bent. Parts may be cleaned using a low pressure, dry, bead blast. Pressure must not exceed 280 kPa (40 psi). Make sure all parts are clean, smooth and free from all deposits. Drilled passages should be cleaned with compressed air.

Inspect bearing bores in center housing (13) for scored surfaces, out-of-round or excessive wear. Make sure that center housing is not grooved where seal (18) rides. Compressor impeller (3) must not show evidence of rubbing against either compressor housing (1) or back plate (4). Make certain that impeller blades are not bent, chipped, cracked or eroded. Oil passages in thrust collar (8) must be clean and thrust faces must not be warped or scored. None of the ring groove shoulders should have step wear. Inspect shroud (15) for evidence of rubbing. Turbine wheel (16) should not show evidence of rubbing and vanes should not be bent, cracked, nicked or eroded. Turbine wheel shaft should not show signs of scoring, scratching or overheating.

Always install new snap rings (10) if removed and check shaft end play and radial clearance when assembling. Position shroud (15) on turbine shaft (16) and install seal ring (18) in groove. Apply a light even coat of engine oil to shaft journals, compress seal (18) with a strong thin tool, such as a dental pick, and slide shaft and seal into center housing. Install new seal ring (7) in groove of thrust collar (8), then install thrust bearing (9) so that smooth side of bearing is toward seal ring end of collar. Install thrust bearing and collar assembly over shaft, making certain that pins in center housing engage holes in thrust bearing. Install new seal ring (12) and install back plate making certain that seal ring (7) is not damaged. Seal ring will be less likely to break if open end of seal ring is installed in back plate first. Install lock plates, retaining plates and screws to retain back plate. Install compressor impeller (3) on shaft, coat threads with engine oil and lock with nut (2). Tighten nut (2) to 2.26 N.m (20 in.-lbs.) torque, then continue to tighten nut using a "T" handle wrench until shaft length is increased 0.14-0.16 mm (0.0055-0.0065 inch). An-

other method of tightening nut (2) is to first tighten to 2.26 N.m (20 in.-lbs.), then turn nut an additional 110°. When nut is tight, shaft end play should be 0.03-0.10 mm (0.001-0.004 inch). Install new seal ring (5) and align compressor housing (1) and center housing (13). Alignment is easier if parts (1, 13 and 17) were marked before disassembly. Screws attaching compressor housing (1) to center housing should be tightened to 14.1 N.m (130 in.-lbs.) torque. Install exhaust housing (17) and tighten retaining screws to 14.1 N.m (130 in.-lbs.) torque. Push ends of compressor shaft while rotating and check for freedom of movement. Add engine oil to center housing and cover openings if unit is not immediately installed on tractor.

DIESEL FUEL SYSTEM

The diesel fuel system consists of three basic units; fuel tank, injection pump and injection nozzles. A transfer pump is included to move the fuel from the tank to the injection pump and filters are located in the system to make sure that the fuel is clean before entering the fuel injection pump.

Cleanliness is of the utmost importance and only approved fuel that is absolutely clean and free from foreign material should be added to the tractor fuel tank. Extra precaution should be taken to make sure that no water enters the fuel storage tanks. Extra care in daily maintenance will extend the service life and reduce costly repairs.

Extremely high precision standards are necessary in the manufacture of and service to diesel fuel injection components. Avoid nicks and burrs on all of the working parts.

FILTERS

All Models

131. MAINTENANCE. The fuel filter head is fitted with either one or two renewable type elements. Filters are equipped with water drain cocks (13—Fig. 113) at the bottom of filter. Drain cocks should be opened every few days or whenever water is visible in the sediment bowl (9). Operate lever on fuel transfer pump to refill filter with fuel. If cam for transfer pump is at point of maximum lift, hand lever may not operate.

Fuel filter (7) should be removed and new filter installed at least every 500 hours of operation. To renew the filter, turn off fuel supply at tank or next to fuel transfer pump. Clean outside of the filter and drain filter by loosening the drain cock. Remove cap screw (1) in top center of the filter head and remove filter (7) and bowl (9). Clean inside of filter head and sediment bowl and make sure that all of gaskets (6, 8 and 10) and "O" rings (3 and 5) are removed. Install new filter using all new gaskets and "O" rings. Refer to paragraph 133 to bleed air from the system.

FUEL LIFT PUMP

All Models

132. The diaphragm type fuel lift (transfer) pump is mounted on the side of engine and is operated by a lobe on the engine camshaft. Several different models of fuel lift pump have been used. A repair kit containing a diaphragm, check valves and gaskets is available for servicing most pump models.

BLEEDING

All Models

133. Air must be bled from the system whenever the fuel tank has run dry, fuel filter has been renewed or any of the fuel lines have been disconnected. To bleed the system, make sure that tank shut-off is open and all fuel line connections are tight. Loosen plug or "thermostart" fuel line from top of filter head at location (B—Fig. 114). Operate priming lever of fuel lift (transfer) pump until fuel without bubbles flows from loosened plug or fitting, then tighten the

Fig. 113—Exploded view of fuel filters typical of all models. Some models may have filter head (4) that is fitted for two filter assemblies.

1. Cap screw	8. Gasket
2. Washer	9. Bowl
3. "O" ring	10. Gasket
4. Filter head	11. Nut
5. "O" ring	12. Washer
6. Gasket	13. Drain cock
7. Filter element	14. Base

Fig. 114—View of fuel filter showing location of bleed point (B). Fitting of filter shown at left is for line to "thermostart" unit.

plug or fitting. If cam for transfer pump is at point of maximum lift, hand lever may not operate. To correct condition, rotate engine crankshaft (and camshaft) until pump operates properly.

Move fuel shut-off control to RUN position and loosen the vent screws (1 and 2—Fig. 115) on side of pump. Actuate priming lever of fuel lift (transfer) pump until fuel free of air bubbles flows from loosened screws. Tighten lower screw (1) first, then upper screw (2). The tractor will run with air in governor housing, but bleeding at screw (2) will remove trapped air which may cause rusting. Operate priming lever and loosen fitting (T—Fig. 116) at "thermostart" unit. Tighten connection when fuel flows from fitting. Continue to operate priming lever for several strokes after tightening plug (2—Fig. 115) to expel air remaining in bleed back lines. Loosen connections (N—Fig. 116), push stop control IN and move throttle to wide open. Crank engine until fuel without bubbles flows from loosened connections (N), then tighten fittings and start engine.

INJECTOR NOZZLES

CAUTION: Fuel leaves the injection nozzle with sufficient force to penetrate the skin. When testing, stay clear of the nozzle spray.

All Models

134. LOCATING A FAULTY INJECTOR. If the engine does not run properly and a faulty injector is suspected, locate the faulty injector as follows: With the engine running, loosen the high pressure connection at each injector in turn, thereby allowing fuel to escape at the union rather then enter the injector. As in checking for malfunctioning spark plugs in a spark ignition engine, the faulty injector is the one which least affects the running of the engine when its line is loosened.

135. INJECTOR TESTING. A complete job of testing and adjusting fuel injectors requires the use of special test equipment. If injector is suspected of faulty operation, remove the injector as outlined in paragraph 140 and test for opening pressure, spray pattern, seat leakage and leak-back.

Operate the tester until oil flows, then connect the injector to the tester. Close tester valve to shut-off the passage to the gauge and operate the tester lever to be sure that injector is in operating condition and that nozzle is not plugged. If oil does not spray from all of the holes in nozzle, if test lever is hard to operate or if other obvious defects are noted, remove the injector from the tester and service as outlined in paragraph

Fig. 115—View of typical fuel injection pump showing bleed plug locations (1 and 2).

Fig. 116—View of typical high pressure connections (N) at injection nozzles and connection at "thermostart" unit (T). refer to text for bleeding instructions.

141. If the injector operates without undue pressure on the lever and fuel is sprayed from all of the holes in nozzle, proceed with the following tests.

136. OPENING PRESSURE. Open the shut-off valve to the gauge and operate the tester lever slowly. Note the pressure indicated on the gauge at which the nozzle spray occurs. Refer to the following specifications which lists correct opening pressure and setting pressure for new injector or serviced injector with new spring.

Tractor Model	Perkins Model	Engine Build Code
340	**AD3.152**	**CE31160**

Injector code - EE
Opening Pressure . . . . 17,720 kPa
(2570 psi)
Setting Pressure (new) . . . . 19,237 kPa
(2790 psi)
Nozzle BDLL150S6602
Holder BKBL67S5299

350	**AD3.152**	**CE31160**

Injector code - EE
Opening Pressure . . . . 17,720 kPa
(2570 psi)
Setting Pressure (new) . . . . 19,237 kPa
(2790 psi)

Fig. 117—Exploded view of C.A.V. injector nozzle and holder assembly typical of type used on some models.

1. Cap nut
2. Gasket
3. Adjusting screw
4. Spring
5. Spindle
6. Nozzle holder
7. Dowel pin
8. Nozzle valve
9. Nozzle body
10. Nozzle nut
11. Copper washer

Nozzle BDLL150S6602
Holder BKBL67S5299

355	**AT3.1524**	**CN31187**

Before U796442P
Injector code - HM
Opening Pressure . . . . 25,305 kPa
(3670 psi)
Setting Pressure (new) . . . . 25,305 kPa
(3670 psi)
Nozzle JB680129
Holder LRB67015

After U796441P
Injector code - HX
Opening Pressure . . . . 21,285 kPa
(3087 psi)
Setting Pressure (new) . . . . 21,285 kPa
(3087 psi)
Nozzle JB6801029
Holder LRB67015

360	**AT3.1524**	**CN31185**

Injector code - HX
Opening Pressure . . . . 21,285 kPa
(3087 psi)
Setting Pressure (new) . . . . 21,285 kPa
(3087 psi)
Nozzle JB6801029
Holder LRB67015

Early		
399	**A6.3544**	**TW31159**

Injector code - HU
Opening Pressure . . . . 22,298 kPa
(3234 psi)
Setting Pressure (new) . . . . 22,298 kPa
(3234 psi)
Pressure is adjusted by changing shims.
Nozzle JB6801022
Body LRB67014

Late		
399	**1006.6**	**YA31240**

Injector - JU
Opening Pressure . . . . 22,298 kPa
(3234 psi)
Setting Pressure (new) . . . . 22,298 kPa
(3234 psi)
Nozzle JB6801106
Holder LRB67014

If the gauge pressure is not within limits, disassemble as required to turn the adjusting screw or change the thickness of shims as required to bring the opening pressure within specified limits. If opening pressure is erratic or can not be properly adjusted, remove the injector from the tester and overhaul injector. Some injectors are different than type shown in Fig. 117 and are not equipped with threaded adjuster (3). Shims are used to adjust the opening pressure of some models. After opening pressure is correct, use a new copper washer and install cap nut.

Recheck opening pressure after tightening cap nut to 70 N·m (50 ft.-lbs.) torque to be sure that opening pressure has not changed. If the opening pressure is within limits, check the spray pattern as outlined in the following paragraph.

137. SPRAY PATTERN. Operate the tester lever slowly and observe the nozzle spray pattern. All of the sprays must be similar and evenly spaced in nearly a horizontal plane. Each spray must be well atomized and should spread to a 76 mm (3 inches) diameter cone approximately 9.5 mm (⅜ inch) from the nozzle tip. If the spray pattern does not meet these conditions, remove the injector from the tester and overhaul. If the spray pattern is satisfactory, proceed with seat leakage test as outlined in the following paragraph.

138. SEAT LEAKAGE. Close the valve to shut-off pressure to the tester gauge and operate the tester lever quickly for several strokes. Wipe the nozzle tip dry with clean blotting paper, open the valve to the tester gauge and press the lever down slowly to bring the gauge pressure to 1035 kPa (150 psi) below the nozzle opening pressure and hold this pressure for 10 seconds. Touch a piece of blotting paper to the nozzle tip. The resulting oil blot should not be larger than 12.7 mm (½ inch) in diameter. If the nozzle tip drips oil or if blot is excessively large, the injector should be overhauled. If nozzle seat leakage is not excessive, proceed with nozzle leak-back test as outlined in the following paragraph.

139. NOZZLE LEAK-BACK. Operate the tester lever to bring gauge pressure to approximately 15,860 kPa (2300 psi), release the lever and note the time that it takes for the gauge to drop from 15,170 kPa (2200 psi) to 10,345 kPa (1500 psi). If the time is less than 5 seconds, the nozzle is worn or there are dirt particles lodged in the nozzle. If the time to drop the pressure is too long (40 seconds or more), the needle may be too tight in the bore.

NOTE: A leaking tester connection, a leaking check valve in tester or leaking test gauge will also indicate excessively fast leak-back. If injectors consistently show excessively fast leak-back, the tester should be suspected as faulty rather than all injectors.

140. REMOVE AND REINSTALL. Carefully clean all dirt and other foreign material from lines, connectors, injectors and cylinder head area around injectors before removing the injectors. Disconnect injector leak-off line at each injector and at fuel return line. Disconnect the injector line at pump and at injector. Cover all openings and lines to prevent the entrance of dirt. Remove the two nuts or screws

retaining the injector and carefully remove the injector from the cylinder head.

Make sure injector seats in cylinder head are clear and free from any carbon deposit before reinstalling injectors. Install a new copper washer in seat and a new dust sealing washer around the body of injector. Insert injector in cylinder head bore, install retaining washers and nuts or screws. Tighten retaining screws or nuts evenly and alternately to 16 N·m (12 ft.-lbs.) torque for all models except late 399 with 1006.6 engine, which should be tightened to 12 N·m (9 ft.-lbs.) torque. Install a new gasket below and above each leak-off banjo fitting, then tighten the banjo bolt to 8 N·m (72 in.-lbs.) torque. Reconnect the leak-off line to the return line. Check fuel injector connections to be sure that they are clean and properly aligned with fittings at injector and at pump. Reconnect injector pressure lines, tightening to 27 N·m (20 ft.-lbs.) torque.

It is suggested that the fuel filter be serviced as described in paragraph 131 each time injectors are serviced. Refer to paragraph 133 for bleeding the fuel system. Since lines to injectors are opened, it will be necessary to crank engine with fitting at injectors loosened to fill the lines. Start and run engine to be sure that injectors are properly sealed and that injector pressure line and leak-off line connections are not leaking.

141. OVERHAUL. Do not attempt to overhaul diesel injectors unless complete and proper equipment is available. Refer to Fig. 117 for exploded view of one type of injector unit.

Secure the injector holding fixture in a vise and locate the injector in the holding fixture. Never clamp the injector body in a vise. Remove the cap nut and back-off the adjusting screw of models so equipped. Lift the upper spring disc, injector spring and lower spring seat from injector. Remove the nozzle retaining nut using an appropriate special socket, then remove nozzle and valve. The nozzle and valve are a lapped fit to each other and parts from one must never be interchanged with parts from another matched unit. Place all parts in clean fuel oil or calibrating fluid as they are disassembled. Clean the injector unit as follows: Soften hard carbon deposits by soaking in a suitable carbon solvent, then use a soft wire (brass) brush to remove carbon from the exterior. Rinse the nozzle and needle immediately after cleaning to prevent carbon solvent from corroding the highly finished surfaces. The pressure chamber of nozzle can be cleaned with a special reamer. Clean spray holes (Fig. 118) with proper size of wire probe held in a pin vise as shown in Fig. 119. To reduce the chance of breakage, wire should protrude from vise only far enough to pass through pin holes. Rotate pin vise, but do not apply undue pressure.

Valve seat in nozzle is cleaned by inserting a special seat scraper into nozzle, then rotating the scraper. The annular groove can be cleaned with another special hooked scraper.

After cleaning, back-flush the nozzle using a back-flush adapter attached to nozzle tester. Rotate the nozzle while operating the lever to flush nozzle.

Seat in nozzle can be polished using a small amount of tallow on the end of a polishing stick. Be sure that carbon is thoroughly flushed from nozzle before polishing seat.

If the leak-back test was longer than 40 seconds as tested in paragraph 139, or if needle is sticking in bore of nozzle, the needle should be lapped to the nozzle. Use special polishing lapping compound such as Bacharach 66-0655. Place the small diameter of nozzle in a drill chuck and rotate at not more than 450 rpm. Apply a small amount of polishing compound on the barrel of valve needle taking care not to allow any compound on tip or beveled portion. Insert the valve needle into the slowly rotating nozzle. Work the needle in and out several times taking care not to put any pressure against seat. Withdraw needle, remove nozzle and back-flush the nozzle and needle assembly.

NOTE: Do not lap valve for longer than 5 seconds at a time and allow parts to cool between lappings.

Fig. 119—Clean spray holes with a pin vise and wire probe as shown.

Assemble nozzle and needle while still wet from cleaning. Valve needle should slide easily in nozzle. If any sticking is noted, be sure that parts are clean and temperature has normalized.

Rinse all parts in clean fuel oil or calibrating fluid before reassembling, then assemble while still wet. Position the nozzle and needle valve on injector body and be sure that dowel pins in body are correctly located in nozzle (Fig. 120). Install the nozzle retaining nut and tighten nut to 70 N·m (50 ft.-lbs.) torque. Install spindle, washer, spring and pressure adjusting screw. Connect the injector to tester and adjust opening pressure as outlined in paragraph 136. Some injectors are different than type shown in Fig. 117 and are not equipped with threaded adjuster (3). Shims are used to adjust the opening pressure of some models. After opening pressure is correct, use a new copper washer and install cap nut. Recheck opening pressure after tightening cap nut to 70 N·m (50 ft.-lbs.) torque to be sure that opening pressure has not changed.

Retest injector as outlined in paragraphs 135 through 139. Install new nozzle and needle assembly

Fig. 118—Holes (arrows) in the injector nozzle are not located an equal distance from the nozzle tip.

Fig. 120—Be sure that lapped surfaces are clean and that locating dowels are perfectly aligned when nozzle body is reinstalled.

or complete injector assembly if still faulty. Nozzles should be thoroughly flushed with calibrating fluid prior to storage.

FUEL INJECTION PUMP

All Models

142. TIMING. Refer to paragraph 50 (three-cylinder models), paragraph 81 (early, A6.3544, six-cylinder models) or paragraph 110 (late, 1006.6, six-cylinder models) for installation and timing of the injection pump drive gear. The pump is internally timed and except for adjusting to compensate for timing gear backlash, timing can be considered correct if mark on pump mounting flange is aligned with mark on the engine as shown in Fig. 121 or Fig. 122. Refer to paragraph 145 for specific application and timing specifications.

143. REMOVE AND REINSTALL. Thoroughly clean the pump, lines and connections in the area around pump. Remove the lines from the pump to the injectors, disconnect the fuel inlet and outlet (return) lines and immediately cover all openings to prevent the entrance of dirt. Disconnect the throttle control rod and the fuel shut-off cable.

The fuel injection pump is mounted horizontally on all except A6.3544 engine models. To remove the horizontally mounted pump, remove the cover plate located in front of injection pump from the timing gear cover. Remove the three screws or one center nut attaching the injection pump drive gear to the pump shaft. Unbolt the injection pump from the engine front plate and withdraw pump. **Be careful not to drop locating pin or drive key when removing pump.** It may be necessary to use a puller to push the pump drive shaft from the gear. The pump drive gear will remain in the timing gear housing and will not change the gear timing. The gear cannot be removed unless cover is removed; however, the engine crankshaft should not be turned with the pump removed.

The pump is mounted vertically on A6.3544 engine used on early 399 tractor models. Pump can be withdrawn vertically after removing the three stud nuts. Pump shaft and drive are equipped with master splines allowing pump to be installed in one position only. Align timing marks on pump flange and adapter as shown in Fig. 121.

To reinstall the fuel injection pump on all three and six-cylinder models except early 399 tractors with A6.3544 engine, reverse the removal procedures. Align the scribed mark on pump body mounting flange with mark on engine timing gear housing as shown in Fig. 122 and tighten pump retaining nuts to 24 N•m (18 ft.-lbs.) torque. Center nut retaining the injection pump gear should be tightened to 80 N•m (59 ft.-lbs.) torque on late six-cylinder engine. Tighten the three screws attaching the injection pump gear to the injection pump shaft to 38 N•m (28 ft.-lbs.) torque. Connect fuel lines, throttle control rod and fuel shut-off cable, then bleed the fuel system as outlined in paragraph 133.

Fig. 121—Timing marks (T) on pump mounting flange and adapter should be aligned when installing pump on A6.3544 engine.

Fig. 122—Timing marks (T) on pump mounting flange and mounting pad should be aligned as shown when installing pumps that are mounted horizontally. A six-cylinder pump is shown; however, three-cylinder models are similar.

INJECTION PUMP DRIVE GEAR

All Models

144. To remove the injection pump drive gear, first remove the timing gear cover or housing as outlined in paragraph 45 (three-cylinder models), paragraph 79 (early six-cylinder models) or paragraph 106 (late six-cylinder models). Remove the rocker arm assembly and the camshaft drive (idler) gear. Damage could result if either camshaft or crankshaft is turned independently from the other unless the rocker arms are removed. On three-cylinder models and late six-cylinder models, remove the three screws or one center screw attaching gear to pump shaft, then remove the gear. It may be necessary to use a puller to pull the gear from the pump drive shaft. The gear is removed from early six-cylinder models before timing gear housing is removed. Refer to paragraph 81 for overhaul of the injection pump and accessory drive used on A6.3544 engine installed in early 399 model tractors.

INJECTION PUMP SPECIFICATIONS

All Models

145. The injection pump is a completely sealed unit. Service of any kind should only be attempted by trained personnel with proper equipment to test, repair and adjust CAV pumps of the type installed. Refer to the following for specific application and timing specifications.

Tractor Model	Perkins Model	Engine Build Code
340	**AD3.152**	**CE31160**

Pump – Standard . . . . . DPA 2643B133CW/6/2470
ISO . . . . DPA CW50L/900/6/2470
Rotation . . . . Clockwise
Snap ring timing mark letter . . . . E
No. 1 Cyl. Outlet letter . . . . W
Engine Checking Angle at TDC . . . . 27°
Fuel Pump Checking Angle . . . . 35°
Static Timing BTDC . . . . 16°
Piston Displacement . . . . 3.1453 mm
(0.1238 inch)

350	**AD3.152**	**CE31160**

Pump – Standard. . . . . . DPA 2643B133CW/6/2470
ISO . . . . DPA CW50L/900/6/2470
Rotation . . . . Clockwise
Snap ring timing mark letter . . . . E

No. 1 Cyl. Outlet letter . . . . W
Engine Checking Angle at TDC . . . . 27°
Fuel Pump Checking Angle . . . . 35°
Static Timing BTDC . . . . 16°
Piston Displacement . . . . 3.1453 mm
(0.1238 inch)

355	**AT3.1524**	**CN31187**
Before U796442P		

Pump – Standard . . . DPA 2643B131BW/6/2370
ISO. . . . . DPA BW55L/900/6/2370
Rotation . . . . Clockwise
Snap ring timing mark letter . . . . E
No. 1 Cyl. Outlet letter . . . . W
Engine Checking Angle at TDC . . . . 27°
Fuel Pump Checking Angle . . . . 35°
Static Timing BTDC . . . . 16°
Piston Displacement . . . . 3.1453 mm
(0.1238 inch)

After U796441P

Pump – Standard . . . DPA 2643B135DW/5/2470
ISO. . . . . DPA DW54L/1100/5/2470
Rotation . . . . Clockwise
Snap ring timing mark letter . . . . E
No. 1 Cyl. Outlet letter . . . . W
Engine Checking Angle at TDC . . . . 26½°
Fuel Pump Checking Angle . . . . 35°
Static Timing BTDC . . . . 17°
Piston Displacement . . . . 3.5460 mm
(0.1396 inch)

360	**AT3.1524**	**CN31185**

Pump – Standard. . . . . . DPA 2643B/132DW/5/2470
ISO . . . . DPA DW61L/1100/5/2470
Rotation . . . . Clockwise
Snap ring timing mark letter . . . . E
No. 1 Cyl. Outlet letter . . . . W
Engine Checking Angle at TDC . . . . 26½°
Fuel Pump Checking Angle . . . . 35°
Static Timing BTDC . . . . 17°
Piston Displacement . . . . 3.5460 mm
(0.1396 inch)

Early 399	**A6.3544**	**TW31159**
Before U793031P		

Pump – Standard . . . DPA 2643D180/TY/3/2420
ISO. . . . . DPA TY51L/900/3/2420
Rotation . . . . Counter-Clockwise
Snap ring timing mark letter . . . . F
No. 1 Cyl. Outlet letter X
Engine Checking Angle at TDC . . . . 153°
Fuel Pump Checking Angle . . . . 141°
Static Timing BTDC . . . . 23°
Piston Displacement . . . . 6.4575 mm
(0.2542 inch)

After U793030P

Pump – Standard . . . DPA 2643D188/TY/5/2310
ISO. . . . . DPA TY5L900/5/2310
Rotation . . . . Counter-Clockwise

Snap ring timing mark letter. . . . . F
No. 1 Cyl. Outlet letter. X
Engine Checking Angle at TDC. . . . . 153°
Fuel Pump Checking Angle . . . . 141.5°
Static Timing BTDC. . . . . 23°
Piston Displacement. . . . . 6.4575 mm
                                        (0.2542 inch)

Late 399   **1006.6**   **YA31240**
Pump - Standard . . . . . DPA 2643D803/UK/5/2310

ISO
Rotation . . . . Clockwise
Governor Spring Position Code 5
No. 1 Cyl. Outlet letter . . . . Y
Engine Checking Angle at TDC . . . . 332°
Fuel Pump Checking Angle . . . . 343°
Static Timing BTDC . . . . 22°
Piston Displacement . . . . 5.9191 mm
                                        (0.2330 inch)

ELECTRICAL SYSTEM

146. A 12 volt, negative grounded electrical system is used on all models.

ALTERNATOR AND REGULATOR

All Models

147. Lucas or Motorola 45, 55 or 70 amp alternators have been used. All models have the regulator built into the alternator.

To disassemble the Motorola alternator, first scribe a line across frame halves (5 and 21—Fig. 123). Remove voltage regulator (25), brush set (23), insulating cover (26), then remove the four through bolts. Carefully pry frame apart, between stator frame (15) and drive end (5). **Be careful not to damage windings of stator.** Unsolder stator and voltage regulator leads from rectifier bridge (11). Remove rectifier bridge, capacitor (13) and stator (15). Remove "O" ring (16) from slip ring end frame (21), then remove D+ terminal screw and spade connector/insulator (24) from slip ring end frame (21).

Fig. 123—Exploded view of Motorola alternator typical of type used on some models. Refer to Fig. 127 for Lucas type.

1. Fan
2. Washer
3. Spacer
4. Woodruff key
5. Drive end frame
6. Bearing
7. Bearing retainer
8. Counterbored spacer
9. Rotor
10. Bearing
11. Rectifier bridge
12. Spacer
13. Capacitor
14. Ground terminal screw
15. Stator
16. "O" ring
19. Insulator
20. Insulator
21. Slip ring end frame
22. Wire
23. Brush set
24. Spade connector
25. Voltage regulator
26. Insulating cover

Remove pulley, fan (1), washer (2), Woodruff key (4) and spacer (3), then tap rotor shaft on block of wood to push it out of bearing in drive end frame (5). Slide counterbore spacer (8) off rotor shaft (9). Press bearing (6) out of drive end frame (5), after removing bearing retainer (7). Pull bearing (10) off rotor shaft.

Original brush set (23) can be reinstalled if brushes protrude 5 mm (3/16 inch) or more from bottom of holder. Brushes should not be oil soaked or cracked. Check for continuity from brush (1—Fig. 124) to terminal (3) and from brush (2) to mounting bracket (MB), using an ohmmeter. Check for open circuit between brushes (1 and 2) and from terminal (3) and mounting bracket (MB).

Rectifier bridge (11—Fig. 123) may be tested with an ohmmeter after unsoldering stator and voltage regulator leads. Connect one ohmmeter lead to ground (G—Fig. 125) and other lead to one of the terminals (T), then note ohmmeter reading. Reverse ohmmeter leads and again note reading. A good diode will read approximately 30 ohms resistance in one test mode and continuity (near zero resistance) with test leads reversed. Repeat test on remaining terminals (T). Repeat test between (B+) and terminals (T), then repeat test between (R and T) terminals. Install new rectifier bridge if any one pair of readings is the same regardless of test lead polarity.

Check bearing surfaces of rotor shaft (9—Fig. 123) for visible wear or scoring. Examine slip ring surface for scoring or wear and rotor winding for overheating or other damage. Check rotor for grounded, shorted or open circuits using an ohmmeter. Touch ohmmeter probes to points (1 and 2—Fig. 126) and to points (1 and 3). A reading near zero indicates a short to ground. Touch ohmmeter probes to slip rings (2 and 3) and observe resistance which should be 4.0-5.2 ohms. If resistance is too high, windings are open. If ohmmeter indicates less than 4.0 ohms, windings are shorted or grounded and new rotor should be installed.

Clean slip rings (2 and 3) with fine crocus cloth while spinning the rotor to avoid flat spots on slip rings.

To reassemble, reverse disassembly procedure. When reconnecting lead wires from voltage regulator (25—Fig. 123), connect yellow wire to D+ terminal and green wire to brush assembly connector.

The Lucas alternator used on some models is shown in Fig. 127. Regulator and brushes (11) can be removed without disassembling unit. Alternator can

Fig. 124—Brush set (23—Fig. 123) used in Motorola alternator. Refer to text for test procedures.

Fig. 125—Test points for rectifier bridge (11—Fig. 123). Refer to text for testing procedure.

Fig. 126—Removed rotor assembly showing test points to be used when checking for grounds (shorts) and opens.

Fig. 127—Exploded view of Lucas alternator typical of type used on some models.

1. Screws
2. Drive end frame
3. Spacer
4. Stator
5. Rotor
6. Rectifier
7. Bearing
8. Insulator
9. Washer
10. Slip ring end frame
11. Regulator
12. Capacitor

Fig. 128—Exploded view of Lucas starter used on some models.

1. Solenoid
2. Brake
3. Plunger
4. Pivot pin
5. Shift lever
6. Dust seal
7. Brush
8. Field coil
9. Field housing
10. Drive end frame
11. Bushing
12. Washers
13. Brush holder & end cover
14. Brush & spring
15. Washers
16. Armature
17. Center plate
18. Drive pinion
19. Pinion stop collar
20. Snap ring

be disassembled after removing the nut, pulley and fan from front, then removing the three screws (1).

STARTER

All Models

148. Both Lucas and Perkins starters have been used. Construction is typical for both models as shown in Fig. 128 and Fig. 129.

CIRCUIT DESCRIPTION

All Models

149. Refer to Fig. 130, Fig. 131, Fig. 132, Fig. 133, Fig. 134, Fig. 135, Fig. 136, Fig. 137, Fig. 138, Fig. 139 and Fig. 140 for wiring diagrams typical of most models. Wiring for some models may be different than shown. All models include 12 volt batteries with the negative terminal grounded. Quick disconnects are provided at all main connections. If trouble is encountered and wiring is suspected, check for loose, corroded, damaged or disconnected connections.

Fig. 129—Exploded view of Perkins starter assembly.

1. Solenoid
2. Spacer washers
3. Plunger
4. Pivot pin
5. Shift lever
6. Dust seal
7. Brush
8. Field coil
10. Drive end frame
11. Bushing
12. End cover
13. Brush holder
15. Washers
16. Armature
17. Center plate
18. Drive pinion
19. Pinion stop collar
20. Snap ring

Color code and legend for Figs. 130-140.

B. Black
BU. Black/Blue
BP. Black/Purple
BR. Black/Red
BS. Black/Slate
BLG. Black/Light Green
BW. Black/White
G. Green
GB. Green/Black
GN. Green/Brown
GO. Green/Orange
GP. Green/Purple
GR. Green/Red
GU. Green/Blue
GW. Green/White
K. Pink
LG. Light Green
LGY. Light Green/Yellow
LU. Light Blue
N. Brown
NB. Brown/Black
NG. Brown/Green
NK. Brown/Pink
NR. Brown/Red
NU. Brown/Blue
O. Orange
P. Purple
PB. Purple/Black
PN. Purple/Brown
R. Red
RB. Red/Black
RU. Red/Blue
RG. Red/Green
RLG. Red/Light Green
RW. Red/White
RY. Red/Yellow
S. Grey
U. Blue
UG. Blue/Green
UK. Blue/Pink
UR. Blue/Red
UW. Blue/White
W. White
WB. White/Black
WG. White/Green
WK. White/Pink
WLG. White/Light Green
WN. White/Brown
WR. White/Red
WS. White/Grey
Y. Yellow
YU. Yellow/Blue

1. Horn
2. Headlights
3. Fuel level sender
4. Solenoid - 4WD Differential lock
5. Switch - Engine oil pressure
6. Sender - Coolant temperature
7. Switch - Air filter
8. Switch - Hydraulic filter temperature

9. Switch - Hydraulic filter pressure
10. Alternator
11. Starter motor
12. Battery
13. Thermostart unit
14. Cigar lighter
15. Switch - Stop light
16. Trailer indicator warning light
17. Direction indicator warning light
18. Switch - Direction indicator
19. Flasher unit - Direction indicator
20. Start switch
21. Main fuse box
22. Thirteen pin plug - Instrument panel
     warning lights
23. Twelve pin plug - Instrument panel
     instruments
24. Switch - Main/dip beam
25. Switch - Headlight
26. Switch - Rear work light
27. Switch - Hazard warning light
28. Warning light dimmer
29. Safety start switch - Transmission
     neutral
30. Horn button
31. Idle speed indicator light
32. Fuse - Main start circuit
33. Speedometer
34. Safety start switch - PTO
35. Switch - Differential lock engage
36. Front fender lights - Right
37. Rear fender lights - Right
38. Rear work light - Right
39. Trailer socket
40. License plate light
41. Switch - Parking brake
42. Switch - 4WD
43. Front fender lights - Left
44. Rear fender lights - Left
45. Ground (−) junction
46. Feed (+) junction
47. IPTO switch
48. Red quick disconnect - Behind engine
49. Blue quick disconnect - Behind engine
50. Black quick disconnect - Behind
     engine
51. Quick disconnect - Right clutch
     housing
52. Quick disconnect - Right side cover
53. Quick disconnect - Right clutch
     housing
54. Quick disconnect - Right side cover
55. Quick disconnect - Left seat deck
56. Quick disconnect - Right seat deck
57. Quick disconnect - Speedometer
58. Quick disconnect - Speedometer
59. Rear work light - Left
60. Front work light - Right
61. Front work light - Left
62. Beacon
63. Interior light (w/cab)

64. Front wiper motor (w/cab)
65. Rear wiper motor
66. Radio
67. Switch - Road Hi-Lo
68. Switch - Rear wiper
69. Switch - Front wiper
70. Switch - Cab heater blower
71. Switch - Beacon
72. Cab heater blower motor
73. Switch - Multi-Power
74. Left flasher
75. Quick disconnect
76. Quick disconnect
77. Right flasher
78. Switch - Front work light
79. Windshield washer pump
80. Cab fuse box
81. Multi-Power solenoid
82. Switch - Reverse horn
83. Reverse horn
84. Starter motor solenoid
85. Road warning lights
86. Switch - Road warning light
87. Flasher - Road warning light
88. Radio speaker
89. Radio antenna
90. Engine ground
91. Air conditioner relay
92. Air conditioner thermostat
93. Air conditioner compressor
94. Switches - Air conditioner
     cut-out
95. Quick disconnect - Air
     conditioner
96. Rear window washer pump
97. Fresh air blower motor
98. Fresh air blower motor resistor
99. Switch - Fresh air blower
100. Line fuse - Air conditioner
101. Fresh air blower motor relay
102. Quick disconnect - Right light
103. Quick disconnect - Left light
104. Quick disconnect - Direction
     indicator
105. Switch - 4WD
106. Range indicator light junction
107. Range indicator light
108. Range indicator light
     switch junction
109. Switch - Range indicator light
110. Quick disconnect - Reverse horn
111. Quick disconnect - Reverse horn
     switch
112. Quick disconnect - Differential
     lock switch
113. Engine speed sensor
114. Fuel valve
115. Quick disconnect - 4WD switch
116. Switch - Battery isolator
117. Tachometer interface unit
118. 4WD solenoid valve

Fig. 130—Wiring diagram typical of all models up to tractor serial number V52267.

Fig. 131—Wiring diagram for models with cab from tractor serial number V52268 to P22199.

Fig. 132—Wiring diagram for models without cab from tractor serial number V52268 to R15000 (three cylinder models) or R14231 (six cylinder models).

Fig. 133—Wiring diagram for tractor cabs from serial number V52268 to P22199. Refer to Fig. 134 for front circuit.

Fig. 134—Wiring diagram for three-cylinder tractors from serial number R15001 to S00000 and six-cylinder tractors from serial number R14232 to S00000 without cab.

Fig. 135—Front wiring diagram for three cylinder tractors from R15196 to S00000 and six-cylinder tractors from serial number R15159 to S00000 with cab. Refer to Fig. 136 for rear circuit.

Fig. 136—Rear wiring diagram for three-cylinder tractors from serial number R15196 to S00000 and six-cylinder tractors from serial number R15159 to S00000 models with cab.

Fig. 137—Front wiring diagram for six-cylinder tractors after serial number S01000.

Fig. 138—Front wiring diagram for three-cylinder models with cab after serial number S01000. Refer to Fig. 134 for rear circuit.

Fig. 139—Wiring diagram typical of all models after serial number S01000 without cab.

Fig. 140—Wiring diagram typical of all models after serial number S01000 with cab.

ENGINE CLUTCH

150. All models are equipped with either a dual or split torque clutch. Some models are equipped with 305 mm (12 inches) diameter clutch and others have a larger 330 mm (13 inches) diameter unit. Other differences will also be noted in the following service and adjustment procedures when necessary. The dual clutch controls engagement of drive to the engine transmission and to the power take off (pto). Split torque clutches are installed on tractors equipped with independent pto. Drive for the hydraulic pump and for the pto are taken from the splined hub of a plate attached to the clutch cover, which turns all the time that engine is running. Control of pto engagement or disengagement is by a separate, hydraulically operated clutch. Refer to the appropriate following paragraphs for service to both types of clutches.

CLUTCH LINKAGE ADJUSTMENT

All Models

151. Clutch linkage adjustment should be carefully monitored at all times, but especially during the first 50 hours of operation with new clutch linings.

On all models except 350 with dual clutch and 350, 355 and 360 models with split torque clutch, pedal height (C—Fig. 141) is set by the upper attachment position of the main link. If tractor is equipped with dual clutch, use lower hole (L) as shown. If equipped with split torque clutch, attach link using upper hole (U).

On 350 tractors with dual clutch, pedal height (C—Fig. 142) is adjusted by stop screw (S). Pedal height when released should be 165 mm (6.5 inches).

On 350, 355 and 360 models with split torque clutch, pedal height (C—Fig. 142) is adjusted by stop screw (S). Pedal height when released should be 120 mm (4.75 inches). Some later tractors may have fixed stop and are not adjustable.

On all models, clutch pedal free play should be measured by pressing pedal down until release lever just contacts clutch release fingers and measure distance (A—Fig. 141 or Fig. 142) pedal has traveled. On models with dual clutch, free travel should be 20-25 mm (¾ - 1 inch) and is adjusted by turning nut (N).

To adjust the pto clutch on models with dual clutch and live pto, remove the cover from underside of the clutch housing and turn engine until one of the three adjusters is visible as shown in Fig. 143. Clearance (C) between head of screw and clutch plate should be 2 mm (0.080 inch). Screw can be threaded in or out as necessary after loosening locknut. Make sure that clearance at all three adjusters is exactly alike.

ENGINE/TRANSMISSION SPLIT

All Models

152. To separate tractor between rear of engine and front of transmission, block both rear wheels to prevent rolling, remove any front mounted equipment and any front weights. Remove hood, grille and side panels. If equipped with a cab heater, drain coolant from radiator and engine block, then disconnect hoses to heater. If equipped with air conditioning, disconnect refrigerant lines. Remove gear lever trim plate and boot from models with cab. On four-wheel-drive tractors, refer to paragraph 9 and remove drive shaft and shield. On all models, disconnect battery ground and front wiring harness to the headlights, horn and fuel tank. Disconnect throttle and stop cables from the fuel injection pump, detach cables from support brackets and move cables out of the way. Disconnect hydraulic lines between rear of tractor and front that would interfere with the removal of engine and cover all openings. If necessary, drain fuel tank. Wedge between the front support and axle to prevent tipping and attach hoist to the engine. Support rear of tractor

Fig. 141—View of clutch pedal and operating linkage typical of all models except 350 with dual clutch and 350, 355 and 360 models with split torque clutch.

under transmission housing, then carefully remove the screws and nuts attaching the engine to transmission housing. Carefully move the engine and front axle assembly forward away from the transmission.

Fig. 142—View of clutch pedal and operating linkage typical of 350 models with dual or split torque clutch and 355 and 360 models with split torque clutch.

Fig. 143—View of one of three adjusters of live pto clutch. Clearance (C) should be 2 mm (0.080 inch) for all three locations.

Reassemble by reversing the splitting procedure and observe the following. Install a M12 alignment stud approximately 100 mm (4 inches) long in each side of the transmission housing to assist in aligning engine to transmission. Turn flywheel to align clutch plate splines with transmission and pto input shaft splines. Install retaining screws and nuts after engine and transmission housing flanges are completely together. Tighten retaining screws and nuts to 115 N·m (85 ft.-lbs.) torque. Reverse removal procedure and refer to paragraph 5 for attaching front support to engine and selecting proper shims. Check and adjust clutch linkage as outlined in paragraph 151. Refer to paragraph 9 for installing drive shaft of four-wheel-drive models.

CLUTCH ASSEMBLY

All Models With Dual Clutch

153. REMOVE AND REINSTALL. To remove the clutch assembly, first refer to paragraph 152 and separate engine from transmission housing.

CAUTION: Install three $^1/_4$ UNC screws 2-$^1/_8$ inches (54 mm) long into the three equi-spaced holes (3—Fig. 144) in clutch cover to hold spring presure and prevent clutch asembly from flying apart. Loosen the six screws (6) retaining the clutch to flywheel, then remove screws and separate clutch from flywheel. Be careful to avoid breathing dust which may contain asbestos. Do not use compressed air to blow dust from clutch and flywheel.

Position main clutch disc and clutch assembly on the flywheel, using alignment tool (MF 159B or equivalent) to center the clutch asembly and disc.

Fig. 144—Three 1/4 UNC screws (3) must be installed as shown to hold spring pressure when removing or installing dual clutch. Six screws (6) attach cover to flywheel.

Install the six retaining screws (6—Fig. 144) and tighten to 27-34 N·m (20-25 ft.-lbs.) torque, then remove the three spring compression screws (3).

Check clearance (C—Fig. 143) between heads of the three pto adjusting screws and plate. Clearance should be 2 mm (0.080 inch) and can be adjusted by turning screw after loosening lock nut. Clearance at all three locations should be exactly the same for even engagement. Use setting gauge (MF 314 or equivalent) to measure height of release fingers which should be 82.31 - 82.55 mm (3.25 - 3.28 inches). Adjust screws in fingers so that distance from back of screw to flywheel is correct and all are exactly the same. Refer to paragraph 152 and reconnect engine to transmission housing. Readjust clutch linkage as outlined in paragraph 151.

154. OVERHAUL. Refer to paragraph 153 and remove unit from tractor. Set clutch assembly in a press and compress assembly until the three screws (3—Fig. 144) are free and can be easily removed. Remove cotter pins (8—Fig. 145), pins (6) and springs (7), then disconnect levers (3) from links (20). Release pressure from press slowly and allow springs (14) to fully extend. Lift cover (2) and Belleville spring (16) from clutch, then lift pto pressure plate (17) and pto clutch disc (18) from clutch. Remove false flywheel (19), springs (14) and fiber washers (13).

Fig. 145—Exploded view of dual clutch typical of models so equipped.

2. Clutch cover	13. Fiber washer
3. Release lever (3 used)	14. Engine clutch spring
4. Locknut (3 used)	15. Pto adjusting screw
5. Screw (3 used)	and locknut (3 used)
6. Pin (3 used)	16. Pto Belleville spring
7. Lever spring (3 used)	17. Pto pressure plate
8. Pin (3 used)	18. Pto friction disc
9. Pin (3 used)	19. False flywheel
10. Pivot pin (3 used)	for pto clutch
11. Engine clutch	20. Connecting links
friction disc	(3 used)
12. Engine clutch	21. Pin (3 used)
pressure plate	22. Pin (3 used)

Check all parts for wear, scoring or evidence of overheating. Friction surfaces of false flywheel (19), pressure plate (17) and pressure plate (12) should not be resurfaced. If damaged, new parts should be installed. Friction surface of engine flywheel can be surfaced, but no more than a total of 1.0 mm (0.040 inch) can be removed. The ledge on flywheel, to which the clutch is attached, should have exactly the same amount removed as the friction surface. Usually only 0.25 mm (0.010 inch) will need to be skimmed from surfaces to restore flywheel. Refer to Fig. 146 for flywheel dimensions (A and D).

Refer to Fig. 145 and the following:
Engine friction disc (11) –
   Diameter . . . . 305 mm (12 inches)
   Thickness . . . . 9.22 mm (0.363 inch)
   Wear limit . . . . 6.72 mm (0.265 inch)
Pto friction disc (18) –
   Diameter . . . . 254 mm (10 inches)
   Thickness . . . . 8.00 mm (0.315 inch)
   Wear limit . . . . 6 mm (0.236 inch)
Engine clutch pressure springs (14) –
340 and 350 Models
   Color . . . . Red
   Number used 6
   Clamp load . . . . 442 kg (975 lbs.)
355 and 360 Models
   Color . . . . Green
   Number used 9
   Clamp load . . . . 749 kg (1650 lbs.)
Pto Belleville spring (16) -
340 and 350 Models
   Color . . . . Dark Blue
   Clamp load . . . . 568 kg (1250 lbs.)
355 and 360 Models
   Color . . . . Dark Green
   Clamp load . . . . 749 kg (1650 lbs.)

Compress clutch in a press while assembling. Springs should be located as shown in Fig. 147. Install three cap screws (3—Fig. 144) to hold clutch assembly together. Refer to paragraph 153 for adjustment and to paragraph 152 for installation and reassembly. Refer to paragraph 151 to adjust linkage.

All Models With Split Torque Clutch

155. REMOVE AND REINSTALL. To remove the clutch assembly, first refer to paragraph 152 and separate engine from transmission housing.

CAUTION: Install three ¹/₄ UNC screws 2-¹/₈ inches (54 mm) long into the three equi-spaced holes (3—Fig. 148) in clutch cover to hold spring pressure and prevent clutch from flying apart. Loosen the six screws (6) retaining the clutch to flywheel, then remove screws and separate clutch from flywheel. Be careful to avoid breathing dust which may contain asbestos. Do not use compressed air to blow dust from clutch and flywheel.

Fig. 146—The same amount of material machined from friction surface of engine flywheel should be removed from ledge to which clutch is attached to maintain distance (A, B or C). Angle of chamfer (D and E) should be 45°. Original contour (F) should be maintained.

A. 39.62-39.75 mm
　(1.560-1.565 inches)

B. 3.3 mm (0.130 inch)

C. 3.3 mm (0.130 inch)

D. 1.6 mm (0.063 inch)
E. 0.5 mm (0.020 inch)

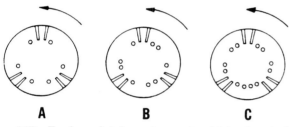

Fig. 147—Engine clutch springs should be located as shown, depending upon whether 6, 9 or 12 springs are used.

Position main clutch disc and clutch assembly on the flywheel, using alignment tool (MF 159B or equivalent) to center the clutch assembly and disc. Install the six retaining screws (6—Fig. 148) and tighten to 27-34 N·m (20-25 ft.-lbs.) torque for 305 mm (12 inches) diameter clutch, or 36-54 N·m (26-40 ft.-lbs.) for 330 mm (13 inches) diameter clutch. Then, remove the three spring compression screws (3).

On models with 305 mm (12 inches) diameter split torque clutch, use setting gauge (MF 314 or equivalent) to measure height of release fingers which should be 82.31-82.55 mm (3.25-3.28 inches). On 399 models with 330 mm (13 inches) diameter clutch, use setting gauge (MF 446 or equivalent) to measure height of release fingers which should be 118.92-120.22 mm (4.682-4.733 inches). On all models, adjust screws in fingers so that distance from back of screw to flywheel is correct and all are exactly the same. Refer to paragraph 152 and reconnect engine to transmission housing. Readjust clutch linkage as outlined in paragraph 151.

156. OVERHAUL. Refer to paragraph 155 and remove unit from tractor. Set clutch assembly in a

Fig. 148—Three 1/4 UNC screws (3) must be installed as shown to hold spring pressure when removing or installing split torque clutch. Six screws (6) attach cover to flywheel.

press and compress assembly until the three screws (3—Fig. 148) are free and can be easily removed. Remove cotter pins from pins (6—Fig. 149 or Fig. 150), then remove pins and springs (7). Release pressure from press slowly and allow springs (14) to fully extend, then lift cover (2) from clutch. Remove springs (14) and washers (13). Hub (18) can be removed if necessary after removing nuts (17) and screws (19).

Check all parts for wear, scoring or evidence of overheating. Friction surface of pressure plate (12) should not be resurfaced. If damaged, new part should be installed. Friction surface of engine fly-

Fig. 149—Exploded view of split torque clutch typical of type used on some models. Refer also to Fig. 150.

2. Clutch cover	12. Engine clutch
3. Release lever (3 used)	pressure plate
4. Locknut (3 used)	13. Washer
5. Screw (3 used)	14. Engine clutch spring
6. Pin (3 used)	17. Nut (3 used)
7. Lever spring (3 used)	18. Hub
8. Pin (3 used)	19. Screw (3 used)
9. Pin (3 used)	20. Connecting links (3 used)
10. Pivot pin (3 used)	21. Pin (3 used)
11. Engine clutch friction disc	22. Pin (3 used)

wheel can be surfaced, but no more than a total of 1.0 mm (0.040 inch) can be removed. The ledge on flywheel, to which the clutch is attached, should have exactly the same amount removed as the friction surface. Usually only 0.25 mm (0.010 inch) will need to be skimmed from surfaces to restore flywheel. Refer to Fig. 146 for flywheel dimensions (B, C and E).

Refer to Fig. 149 or Fig. 150 and the following:
Engine friction disc (11) - Except 399 Model
Diameter . . . . 305 mm (12 inches)
Thickness . . . . 9.22 mm (0.363 inch)
Wear limit . . . . 6.72 mm (0.265 inch)
Engine friction disc (11) – 399 Model
Diameter . . . . 330 mm (13 inches)
Thickness . . . . 9.22 mm (0.363 inch)
Wear limit . . . . 6.72 mm (0.265 inch)
Engine clutch pressure springs (14) With Fiber (Not Cerametalic) Lining –
340 and 350 Models
Color . . . . Red
Number used 6
Clamp load . . . . 442 kg (975 lbs.)

Fig. 150—Exploded view of split torque clutch typical of type used on some models. Refer to text for specifications of all models. Refer to Fig. 149 for legend.

355 and 360 Models
Color . . . . Green
Number used 9
Clamp load . . . . 749 kg (1650 lbs.)
399 Model
Color . . . . Green
Number used 12
Clamp load . . . . 998 kg (2200 lbs.)
Engine clutch pressure springs (14) With Five Pad Cerametalic Lining –
305 mm (12 inches) Diameter
Color . . . . Red
Number used 9
Clamp load . . . . 664 kg (1464 lbs.)
330 mm (13 inches) Diameter
Color . . . . Red
Number used 12
Clamp load . . . . 885 kg (1950 lbs.)
Compress clutch in a press while assembling. Springs should be located as shown in Fig. 147. Install three cap screws (3—Fig. 148) to hold clutch assembly together. Tighten screws (19—Fig. 149 or Fig. 150) to 34-41 N·m (25-30 ft.-lbs.) torque. Refer to paragraph 155 for installation and adjustment. Refer to paragraph 151 to adjust linkage.

TRANSMISSION
(EIGHT-SPEED STANDARD)

350, 355 and 360 Models

Models 350, 355 and 360 are available with a transmission providing eight forward, non-synchronized speeds and two reverse speeds. The eight forward and two reverse speeds are possible by using a standard, sliding gear gearbox with four forward and one reverse speed compounded by a two speed planetary unit. The main gearbox of some models is equipped with overdrive in place of the standard 3rd gear. Refer to Fig. 151 for shift patterns of both standard sliding gear eight-speed transmission and overdrive models.

LUBRICATION

Models With Eight-Speed Standard Transmission

157. The transmission, rear differential and hydraulic system share common fluid which is contained in the transmission and center housing. Recommended lubricant is Massey-Ferguson Super 500 multi-use oil, Massey-Ferguson Permatran or equivalent transmission/hydraulic oil. Capacity is 47.4 L (12.5 gal.) if equipped with spacer or four-wheel-drive transfer box between transmission and rear axle center housing; 43.4 L (11.5 gal.) if not equipped with transfer box or spacer. Tractors are equipped with dipstick (Fig. 152 or Fig. 153) with marks indicating maximum and minimum oil levels. Level should be maintained at maximum level if tractor is operated on hilly terrain or if using implements which require large quantities of oil.

TRACTOR REAR SPLIT

Models With Eight-Speed Standard Transmission Without Cab

158. To separate tractor between the rear of the transmission housing and the front of the rear axle center housing, proceed as follows:

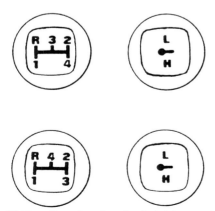

Fig. 151—Shift pattern for standard eight-speed transmission is shown in top view; shift pattern for eight-speed with overdrive in fourth speed of the main transmission is shown in lower view.

Fig. 152—Transmission oil level of tractors without cab is measured by dipstick located in the side cover as shown. Refer to Fig. 153 for models with cab.

Fig. 153—Transmission oil level of tractors with cab is measured by dipstick that points toward rear as shown.

Set parking brake and block the rear wheels to prevent rolling. Disconnect battery ground cable and drain all transmission fluid. If equipped with four-wheel-drive, refer to paragraph 9 and remove the front drive shaft. On all models, remove lower (gearbox) cover, both foot steps and battery boxes. Detach foot throttle, four-wheel-drive indicator switch and four-wheel-drive selector lever. Disconnect hydraulic oil supply pipe located on right side of tractor and, if so equipped, detach hydraulic pipes from left side. Disconnect brake lines and drain system. Disconnect wiring harness at the connectors below console on right side and disconnect wires from safety start switch. Remove the hydraulic suction filter assembly. Place hardwood wedges between the front axle and axle support casting on both sides to prevent tipping around axle pivot when tractor is separated. Support rear of tractor under axle center housing and front of tractor under transmission housing. Remove screws attaching rear axle center housing or four-wheel-drive transfer case to the transmission housing, then separate the tractor halves. Either the front can be moved forward or the rear can be moved to the rear, but it is important to support both halves safely and securely.

If the split pin is removed from the shear tube, install pin in center of the five holes. Install shear tube on pinion. Install at least two dowel studs to facilitate alignment, position new gasket on dowel studs and position rear drive shaft into high/low gear coupler. Move the high/low shifter to "low" and shift main transmission to 3rd. Move tractor together, turning engine to align splines of shear tube. Make sure that flanges are tight against each other before installing retaining screws. Install and tighten one screw on each side, then check end float of the rear drive shaft. Push the shear tube toward rear as shown at inset in Fig. 154, which will compress spring, then measure clearance between the shear tube and locking collar. Correct clearance is 0.4-2.5 mm (0.015-0.100 inch) and can be changed by separating tractor and moving the split ring to another hole in the shear tube. Tighten all screws retaining the four-wheel-drive transfer case, spacer housing or rear axle center housing to the transmission housing to 112 N·m (83 ft.-lbs.) torque, beginning at top center and progressing in a clockwise direction (as viewed from tractor rear) two times around the mating flange. If equipped with four-wheel drive, coat threads of drive shaft retaining screws with "Loctite 270," then attach drive shaft flanges and tighten screws to 55-75 N·m (40-55 ft.-lbs.) torque. Remainder of assembly is reverse of disassembly.

Models With Eight-Speed Standard Transmission
With Cab

159. To separate tractor between the rear of the transmission housing and front of the rear axle center housing, proceed as follows:

Set parking brake, block the rear wheels to prevent rolling and drain all transmission fluid. Disconnect battery ground cable, then disconnect wiring harness at the connectors. If equipped with four-wheel drive, refer to paragraph 9 and remove the front drive shaft. On all models, disconnect throttle stop cable and throttle cable from fuel injection pump and move cables out of the way so that cables will not catch when separating tractor. Disconnect and cap all hydraulic pipes running from front to rear, which would prevent separation of tractor halves. Mark (identify), disconnect and cap steering hoses, heater hoses and air conditioning lines (if so equipped), which would interfere. Place hardwood wedges between the front axle and axle support casting on both side to prevent tipping around axle pivot when tractor is separated. Remove front floor mat and floor plates, disconnect clutch linkage and detach gearbox oil filler pipe. Unbolt and remove the gear shift cover complete with shift levers, then unbolt and remove the transmission cover. Disconnect wires at safety start switch and four-wheel-drive indicator switch. Remove the oil

Fig. 154—View of tractor with cab separated between the four-wheel-drive transfer case and the rear axle center housing. Special splitting stand and track is shown. Support and splitting procedures are similar for models without cab and without four-wheel drive.

feed pipe and filter housing from the right side of transmission case. Disconnect four-wheel-drive selector lever, then check to make sure that all wiring, cables, linkages and hoses are free to permit the tractor to be separated. On model 340 tractors, disconnect the hydraulic pipe from inside the transmission housing which passes from the rear axle center housing to the clutch housing. On all models, support rear of tractor under axle center housing and at front corners of cab. Support front of tractor under transmission housing in such a way that front of tractor, engine, transmission and transfer case can be rolled forward (Fig. 154). Unbolt and remove both cab support brackets from front of cab. Remove screws attaching transfer case to the rear axle center housing, then move front of tractor forward. It is important to support both halves safely and securely to prevent damage and injury.

If the split pin is removed from the shear tube, install pin in center of the five holes. Install shear tube on pinion. Install at least two dowel studs to facilitate alignment, position new gasket on dowel studs and position rear drive shaft into high/low gear coupler. Move the high/low shifter to Low and shift main transmission to 3rd. Move front half of tractor toward rear turning engine to align splines of shear tube. Make sure that flanges are tight against each other before installing retaining screws. Install and tighten one screw on each side, then check end float of the rear drive shaft. Push the shear tube toward rear as shown at inset in Fig. 154, which will compress spring, then measure clearance between the shear tube and locking collar. Correct clearance is

0.4-2.5 mm (0.015-0.100 inch) and can be changed by separating tractor and moving the split ring to another hole in the shear tube. Tighten all screws retaining the transmission housing to spacer housing, transfer case or rear axle center housing to 112 N·m (83 ft.-lbs.) torque, beginning at top center and progressing in a clockwise direction (as viewed from tractor rear) two times around the mating flange. On four-wheel-drive models, coat threads of drive shaft retaining screws with "Loctite 270," then attach drive shaft flanges and tighten screws to 55-75 N·m (40-55 ft.-lbs.) torque. Remainder of assembly is reverse of disassembly.

TRANSMISSION REMOVAL

All Models With Eight-Speed Standard Transmission

160. Refer to paragraph 158 or paragraph 159 to separate transmission housing from rear axle center housing. If not equipped with cab, unbolt and support or remove the steering and instrument console. On all models, support the engine and transmission housing separately. Remove screws and stud nuts, then separate transmission from engine.

Reassemble transmission to engine by reversing the splitting procedure and observe the following. Install a M12 alignment stud approximately 100 mm (4 inches) long in each side of the transmission housing to assist in aligning engine to transmission. Turn flywheel to align clutch plate splines with transmission and pto input shaft splines. Install retaining screws and nuts after engine and transmission housing flanges are completely together. Tighten retaining screws and nuts to 115 N·m (85 ft.-lbs.) torque.

Refer to paragraph 158 or 159 and reconnect transmission to rear axle center housing. Tighten all screws retaining the transmission housing to spacer housing, transfer case or rear axle center housing to 112 N·m (83 ft.-lbs.) torque, beginning at top center and progressing in a clockwise direction (as viewed from tractor rear) two times around the mating flange. On four-wheel-drive models, coat threads of drive shaft retaining screws with "Loctite 270," then attach drive shaft flanges and tighten screws to 55-75 N·m (40-55 ft.-lbs.) torque. On all models, check and adjust clutch linkage as outlined in paragraph 151.

TOP COVER AND SHIFT LEVERS

Models With Eight-Speed Standard Transmission

161. The shift levers and the small shift tower can be removed after first removing floor mat and access panels. Installation of the shift levers and tower is easier if transmission is shifted into neutral before removing. Shift tower (11—Fig. 155) is attached to the top cover (18) by five screws.

The main shift lever (6) can be removed after removing the tower assembly, cover (3), snap ring (9), spring (8) and pin (10). The high/low shift lever can be removed after removing cover (14) and pin (16).

The transmission top cover (18) can be unbolted and lifted from transmission after first removing the shift levers and tower assembly. On models with cab, it is necessary to split rear of transmission from rear axle center housing as outlined in paragraph 159 or remove the cab assembly. Models without cab must have instrument console and other interfering parts removed before transmission top cover. The reason for removing top cover will determine what additional procedures are necessary.

"Loctite 515 Instant Gasket" or equivalent should be used to seal tower to top cover. Tighten retaining screws to 50-70 N·m (37-52 ft.-lbs.) torque.

SHIFTER RAILS AND FORKS

Models With Eight-Speed Standard Transmission

162. To remove the shifter rails and forks, first refer to paragraph 158 or paragraph 159 and separate tractor between rear of transmission and front of axle center housing. Refer to paragraph 161 and remove the shift lever and tower assembly and transmission top cover. Models without cab must have instrument

Fig. 155—Exploded view of transmission top cover, tower cover and shift levers typical of models with standard eight-speed transmission.

1. Panel	10. Pin
2. Boot	11. Tower cover
3. Cover	12. Fill plug
4. Retainer clip	13. High/low shift lever
5. Retaining nut	14. Cover
6. Main shift lever	15. Retaining nut
7. Sleeve	16. Pin
8. Spring	17. Sleeve
9. Snap ring	18. Top cover

and steering console removed before top cover can be unbolted and removed.

On all models, remove the neutral safety start switch (N—Fig. 157). Remove safety wire and set screws (5—Fig. 156) and lift detent springs and plungers (4) from bores in housing. Unbolt and remove interlock parts (12, 13, 14 and 15—Fig. 157). Slide rail out of housing toward rear and remove gate and shift fork through top. Twist rails if necessary while withdrawing.

When assembling, lubricate rails and make sure that rails do not have burrs, especially around the set screw locations, that would prevent easy and smooth installation. Each of the four shift rails is different from the other rails. Tighten screws retaining the interlock retainer (15—Fig. 156) to 40-47 N·m (30-35 ft.-lbs.) torque and set screws (5) to 34-52 N·m (25-38 ft.-lbs.) torque. Install safety wire through set screws

Fig. 156—Exploded view of shift rails and forks for standard eight-speed transmission. Because of gear differences, shift rail (9) is for fourth speed and rail (16) is for second and third speeds of models with overdrive.

1. Shift rail (High/Low)	10. Shift fork
2. Shift gate	11. Shift gate
3. Shift fork	12. Plate
4. Detent springs	13. Gasket
and plungers	14. Interlock plunger
5. Retaining screws	& balls
6. Shift rail (Rev./1st)	15. Retainer
7. Shift gate	16. Shift rail (2nd/4th)
8. Shift fork	17. Shift gate
9. Shift rail (3rd)	18. Shift fork

(5) after tightening to prevent loosening. Shift all gears, rails and gates to neutral after checking to make sure that each gear can be correctly engaged. Install detent plungers and springs (4). Remainder of assembly is the reverse of disassembly procedure.

PLANETARY UNIT

Models With Eight-Speed Standard Transmission

163. To remove the planetary unit, first refer to paragraph 158 or paragraph 159 and separate tractor between rear of transmission and front of axle center housing. Remove set screw (5—Fig. 156) from fork (3), then remove the fork and coupler from rear of planetary. Remove the four retaining cap screws and with-

Fig. 157—Drawing of shift rails, installed. Notice position of notches in plate (12) and location of neutral start switch (N). Plunger of interlock (14) is located in rail (9) and balls are located in retainer (15). Refer to Fig. 156 for legend.

draw rear cover (13—Fig. 158), rear thrust ring (6) and planet carrier (7). Pry planetary ring gear (3) with dowels from rear of housing. Remove planetary front cover (2) and shim (1).

To disassemble planet carrier (7), press planet pins (8) forward out of carrier. Remove planet gears (10) with needle rollers (12) and thrust washers (9). Splined sleeve (15) is retained in hub of carrier (7) by retaining ring (16).

When reassembling, be sure to account for all of the needle rollers (12). Each pinion contains two paths of rollers separated by a washer (11) and each roller path contains 27 rollers (12). Use petroleum jelly (not heavy grease) to hold rollers in place while pressing planet pins into carrier.

Be sure that front plate (2) and rear plate (13) are positioned with grooved side toward planet carrier. Slots in front plate and shim (1) must be aligned and toward top when installed on transmission. Use pe-

Fig. 158—Exploded view of high/low range planetary unit used on 350, 355 and 360 models.

1. Front shim
2. Front plate
3. Ring gear
4. Dowel
6. Thrust washer
7. Planet carrier
8. Pinion shaft
9. Thrust washer
10. Pinion
11. Spacer washer
12. Needle rollers
    (27/path)
13. Rear plate
14. Shift coupler
15. Splined sleeve
16. Retaining ring
17. Transmission mainshaft

troleum jelly to hold thrust rings (6) in place. Make certain that tangs of thrust rings engage notches in pinion carrier and that brass side faces away from carrier. Cutaway section of rear plate (13) should be positioned at the lower, **right** corner of planetary as shown. A lockwasher is used only on mounting screw at corner of cutaway. Tighten mounting screws to 40-47 N·m (30-35 ft.-lbs.) torque. Complete assembly procedure by reversing disassembly.

PTO INPUT SHAFT AND RETAINER HOUSING

Models With Eight-Speed Standard Transmission

164. To remove the pto input shaft (8—Fig. 159) and retainer (3), first split tractor between engine and transmission as outlined in paragraph 152. Remove clutch release bearing (16), retainer (15), fork (13) and release shaft (12). Remove brake cross shaft, if so equipped. Remove screws and pull pto input shaft (8) and retainer (3) as a unit from transmission housing.

To disassemble, remove snap ring (6) from rear of retainer, then push input shaft and bearing (5) out of housing toward rear. Bearing can be removed from input shaft after removing snap ring (4). Remove oil seals (1 and 9) and needle bearing (2) only if new parts are to be installed.

Lubricate all seals and bearings with petroleum jelly before assembling. Use special tool (MF 315A or equivalent) to install new needle bearing (2). Install seal (9) with lip toward rear, using special tool (MF 331 or equivalent). Install ball bearing (5) on shaft with shield toward rear, then install snap ring (4). Use special tool (MF255B or equivalent) to install seal (1) with lip toward rear. Use special sleeve (KMF 1004 or equivalent) to protect seal (1), slide input shaft into the retainer housing, then install snap ring (6). Install "O" ring (7) and install housing and shaft assembly over transmission input shaft (11) into housing bore. Coat threads of retaining screws with "Hylomar" or equivalent sealer and tighten to 54-61 N·m (40-45 ft.-lbs.) torque. Remainder of assembly is reverse of disassembly.

Fig. 159—Pto input shaft (8) and retainer housing (3) can be removed as a unit from front of transmission housing. Transmission input shaft (11) is removed rearward from transmission housing.

1. Seal
2. Needle bearing
3. Retainer housing
4. Snap ring
5. Ball bearing
6. Snap ring
7. "O" ring
8. Pto input shaft
9. Oil seal
10. Thrust washer
11. Transmission input shaft
12. Clutch lever and shaft
13. Clutch release lever
14. Return spring
15. Release housing
16. Release bearing

PTO LOWER SHAFT FRONT BEARING AND RETAINER

Models With Eight-Speed Standard Transmission

165. The pto lower shaft front bearing (41—Fig. 160) can be removed and installed after separating tractor between the rear of the engine and front of transmission housing. Removal of the lower driven gear (45) and shaft (46) require the additional steps of separating at rear of transmission housing and removal of the transmission top cover and shift rails. The following procedure describes only service to the front bearing and bearing retainer (42) after splitting as outlined in paragraph 152.

Remove clutch release bearing (16—Fig. 159), retainer (15), fork (13) and release shaft (12). Unbolt and remove cover (36—Fig. 160), then remove snap ring (38) and washer (39). Thread a ⅜ inch UNC cap screw, 75 mm (3 inches) long into each of the two threaded holes of retainer (42). Tighten the two jackscrews to push retainer (42) and bearing (41) forward off shaft (46). Bearing can be pressed from retainer after removing snap ring (40). Remove "O" ring (37) and gasket (43), then clean all parts, thoroughly.

Coat gasket (43) and threads of retaining screws lightly with "Hylomar" or equivalent sealer before installing. End of shaft (46) is threaded to accept a puller screw to pull bearing (41) onto shaft until washer (39) and snap ring (38) can be installed. Align holes in retainer (42) with holes in transmission housing, install "O" ring (37) and cover (36). Tighten retaining screws to 54-61 N•m (40-45 ft.-lbs.) torque. Remainder of assembly is reverse of disassembly procedure.

TRANSMISSION INPUT SHAFT

Models With Eight-Speed Standard Transmission

166. To remove the transmission input shaft (11—Fig. 159), first remove the transmission as outlined in paragraph 160. Remove the top cover and shift levers as outlined in paragraph 161, planetary unit as outlined in paragraph 163 and the shift rails and forks as outlined in paragraph 162. Refer to paragraph 165 and remove the pto lower shaft front bearing and retainer. Remove snap ring (44—Fig. 160), pull shaft (46) to rear and allow gear (45) to fall to bottom of housing. Move input shaft (11—Fig. 161) and gear (13) forward, then withdraw gear (13) out top. Transmission input shaft (11) can be removed from same top opening after gear (13) is out.

Reinstall input shaft by reversing removal procedures.

MAINSHAFT (OUTPUT SHAFT) AND GEARS

Models With Eight-Speed Standard Transmission

167. To remove the mainshaft (19—Fig. 161 and Fig. 162), first remove the transmission as outlined in paragraph 160. Remove the top cover and shift levers as outlined in paragraph 161, planetary unit as outlined in paragraph 163 and the shift rails and forks as outlined in paragraph 162. Refer to paragraph 164 and remove the pto input shaft and retainer, remove the pto lower shaft front bearing and retainer as outlined in paragraph 165, then remove

Fig. 160—Exploded view of the pto lower shaft and bearings.

36. Cover
37. "O" ring
38. Snap ring
39. Washer
40. Snap ring
41. Ball bearing
42. Retainer
43. Gasket
44. Snap ring
45. Pto lower gear
46. Pto lower shaft

Fig. 161—Exploded view of main-shaft, countershaft and gears. All transmission input shafts (11) do not have a separate gear (11G) located on splines. Refer to Fig. 159 for pto input shaft and Fig. 160 for pto lower shaft. Some transmissions have different ratio gears at (13 and 33), which makes this gear selection faster than gears (21 and 28).

  11. Transmission input shaft
11G. Separate gear
  12. Spacer
  13. Gear (3rd)
  14. Snap ring
  15. Ball bearing
  16. Snap ring
  17. Spacer
  18. Roller bearing
  19. Mainshaft
  20. Gear (1st & Reverse)
  21. Gear (2nd & 4th)
  22. Ball bearing
  23. Snap rings
  25. Snap ring
  26. Ball bearing
  27. Gear (2nd)
  28. Gear (4th)
  29. Snap ring
  30. Roller bearing
  31. Countershaft & 1st gear
  32. Ball bearing
  33. Gear (3rd)
  34. Constant mesh gear
  35. Snap ring
  50. Thrust washers
  51. Washers
  52. Center spacer
  53. Needle rollers
  54. Gear (Reverse idler)
  56. Spacer
  57. Idler shaft
  58. Retaining clip

the transmission input shaft as outlined in paragraph 166.

Remove spacer (12—Fig. 161), drive the mainshaft (19) to rear and remove gear (13). Remove snap ring (14) and continue to drive mainshaft toward rear out of bearing (15). Remove snap ring (16) and withdraw mainshaft toward rear while lifting gears (20 and 21) out top opening. Rear bearing (22) can be removed if necessary after removing snap rings (23).

Reassemble by reversing removal procedure. Refer to Fig. 162 for cross section of assembled transmission. Coat gasket (43—Fig. 160) lightly with "Hylomar" or equivalent sealer before installing. Coat threads of screws attaching cover (36) and screws attaching retainer housing (3—Fig. 159) lightly with "Hylomar" or equivalent sealer before tightening to 54-61 N·m (40-45 ft.-lbs.) torque.

COUNTERSHAFT (LAYSHAFT) AND GEARS

Models With Eight-Speed Standard Transmission

168. The mainshaft must be removed as outlined in paragraph 167 before removing countershaft (31—Fig. 161). Remove snap ring (25) from rear, bump countershaft forward enough to remove snap ring (29) from groove and move snap ring forward to unsplined part of countershaft. Remove snap ring (35), then remove gears (33 and 34) while driving countershaft toward rear. After gears (33 and 34) are lifted from housing, drive countershaft toward front while removing gears (27 and 28).

Reassemble by reversing removal procedure. Snap ring (29) should be around shaft in the unsplined area

Fig. 162—Cross section of eight-speed standard transmission.

and front bearing (32) should be in place on shaft before installing shaft in housing. Refer to Fig. 162 for cross section of assembled transmission.

REVERSE IDLER GEAR AND SHAFT

Models With Eight-Speed Standard Transmission

169. A dummy shaft should be fabricated to hold parts in place while both removing and installing the reverse idler shaft. Dummy shaft should be 25 mm (1 inch) diameter and 55 mm ($2^3/16$ inches) long.

To remove the reverse idler, first remove the mainshaft and countershaft assemblies as outlined in paragraphs 167 and 168. Release locking washer, then remove locking screw and clip (58—Fig. 161). Slide dummy shaft in from front, pushing reverse

idler shaft (57) out toward rear. Dummy shaft should be used to hold needle rollers and washers in place while both removing and installing the reverse idler. Be sure to remove all 56 of the needle rollers.

Assemble needle rollers, washers and spacer in reverse idler using petroleum jelly and the dummy shaft. Assemble center spacer (52), one path of 28 needle rollers (53), one end washer (51) and one thrust washer (50), around dummy shaft, in one end of reverse idler gear. Assemble the second path of 28 needle rollers, the other end washer (51) and the remaining thrust washer (50). Center the dummy shaft so that it holds both thrust washers in place, then position assembly in housing and install idler shaft (57). Install clip (58), lock plate and retaining screw. Tighten retaining screw and lock by bending tab of lock plate against flat of screw. Complete reassembly by reversing disassembly procedure.

TRANSMISSION
(EIGHT-SPEED SYNCHROMESH)

340, 350, 355 and 360 Models

Models 340, 350, 355 and 360 are available with a transmission providing eight forward speeds and two reverse speeds. The eight forward and two reverse speeds are possible by using a main gearbox with four forward and one reverse speed compounded by a two speed planetary unit. The main gearbox is equipped with synchronized shifting in top two speeds. Refer to Fig. 163 for shift pattern of eight-speed synchronized transmission models.

LUBRICATION

Models With Eight-Speed Synchromesh Transmission

170. The transmission, rear differential and hydraulic system share common fluid which is contained in the transmission and center housing. Recommended lubricant is Massey-Ferguson Super 500 multi-use 10W-30, Massey-Ferguson Permatran or equivalent. Capacity is 47.4 L (12.5 gal.) if equipped with spacer or four-wheel-drive transfer box between transmission and rear axle center housing; 43.4 L (11.5 gal.) if not equipped with transfer box or spacer. Tractors are equipped with dipstick (Fig. 164 or Fig. 165) with marks indicating maximum and minimum oil levels. Level should be maintained at maximum level if tractor is operated on hilly terrain or if using implements which require large quantities of oil.

TRACTOR REAR SPLIT

Models With Eight-Speed Synchromesh Transmission

171. To separate the rear of the transmission from the rear axle center housing (Fig. 166), refer to para-

graph 158 if tractor does not have cab or to paragraph 159 if equipped with cab.

TRANSMISSION REMOVAL

Models With Eight-Speed Synchromesh Transmission

172. To remove the transmission, first separate the rear of the transmission from rear axle center hous-

Fig. 164—Transmission oil level of tractors without cab is measured by dipstick located in the side cover as shown. Refer to Fig. 165 for models with cab.

Fig. 163—Shift pattern for eight-speed synchronized transmission. Only third and fourth speeds of the main transmission are synchronized.

Fig. 165—Transmission oil level of tractors with cab is measured by dipstick that points toward rear as shown.

ing as outlined in paragraph 158 if tractor does not have cab or paragraph 159 if equipped with cab. If not equipped with cab, unbolt and support or remove the steering and instrument console. On all models, support the engine and transmission housing separately. Remove screws and stud nuts, then separate transmission from engine.

Reassemble transmission to engine by reversing the splitting procedure and observe the following. Install a M12 alignment stud approximately 100 mm (4 inches) long in each side of the transmission housing to assist in aligning engine to transmission. Turn flywheel to align clutch plate splines with transmission and pto input shaft splines. Install retaining screws and nuts after engine and transmission housing flanges are completely together. Tighten retaining screws and nuts to 115 N·m (85 ft.-lbs.) torque.

Refer to paragraph 158 or 159 and reconnect transmission to rear axle center housing. Tighten all screws retaining the transmission housing to spacer housing, transfer case or rear axle center housing to 112 N·m (83 ft.-lbs.) torque, beginning at top center and progressing in a clockwise direction (as viewed from tractor rear) two times around the mating flange. On four-wheel-drive models, coat threads of drive shaft retaining screws with "Loctite 270," then attach drive shaft flanges and tighten screws to 55-75 N·m (40-55 ft.-lbs.) torque. On all models, check and adjust clutch linkage as outlined in paragraph 151.

Fig. 166—View of tractor with cab separated between the four-wheel-drive transfer case and the rear axle center housing. Special splitting stand and track is shown. Support and splitting procedures are similar for models without cab and without four-wheel drive.

TOP COVER

Models With Eight-Speed Synchromesh Transmission

173. The shift levers and the small shift tower can be removed after first removing floor mat and access panels. Installation of the shift levers and tower is easier if transmission is shifted into neutral before removing. Shift tower (11—Fig. 167) is attached to the top cover (18) by five screws.

The main shift lever (6) can be removed after removing the tower assembly, cover (3), snap ring (9), spring (8) and pin (10). The high/low shift lever can be removed after removing cover (14) and pin (16).

The transmission top cover (18) can be unbolted and lifted from transmission after first removing the shift levers and tower assembly. On models with cab,

it is necessary to split rear of transmission from rear axle center housing as outlined in paragraph 159 or remove the cab assembly. Models without cab must have instrument console and other interfering parts removed before transmission top cover can be removed. The reason for removing top cover will determine what additional procedures are necessary.

"Loctite 515 Instant Gasket" or equivalent should be used to seal tower to top cover. Tighten retaining screws to 50-70 N·m (37-52 ft.-lbs.) torque.

SHIFTER RAILS AND FORKS

Models With Eight-Speed Synchromesh Transmission

174. To remove the shifter rails and forks, first refer to paragraph 158 or paragraph 159 and separate tractor between rear of transmission and front of axle center housing. Refer to paragraph 173 and remove the shift lever and tower assembly and transmission top cover. Models without cab must have instrument and steering console removed before top cover can be unbolted and removed.

On all models, remove the neutral safety start switch (N—Fig. 169). Remove safety wire and set screws (5—Fig. 168) and lift detent springs and plungers (4) from bores in housing. Unbolt and remove interlock parts (12, 13, 14 and 15—Fig. 169). Slide rail out of housing toward rear and remove gate and shift fork through top. Twist rails if necessary while withdrawing.

When assembling, lubricate rails and make sure that rails do not have burrs, especially around the set screw locations, that would prevent easy and smooth installation. Each of the four shift rails is different from the other rails. Tighten screws retaining the interlock retainer (15—Fig. 168) to 40-47 N·m (30-35 ft.-lbs.) torque and the seven set screws (5) to 34-52 N·m (25-38 ft.-lbs.) torque. Install safety wire through set screws (5) after tightening to prevent loosening. Adjust position of shift fork (18) as follows.

Use special tools (MF 414/2 and MF 414/3 or equivalent) to hold shift rail (9—Fig. 168 and Fig. 169) in neutral detent position. Insert centralizing tool (MF 414/1 or equivalent) into hole (H—Fig. 168) and into hole of sliding synchronizing coupling. Tighten each of the two set screws (5S), while turning centralizing tool, to hold the fork in this center position. Screws (5S) should also be tightened to 34-52 N·m (25-38 ft.-lbs.) torque and locked in position with safety wire. Shift all gears, rails and gates to neutral after checking to make sure that shift gates all align. Shift into each gear to make sure that each can be correctly engaged. Install detent plungers and springs (4—Fig. 169). Remainder of assembly is the reverse of disassembly procedure.

Fig. 167—Exploded view of transmission top cover, tower cover and shift levers typical of models with eight-speed transmission.

1. Panel
2. Boot
3. Cover
4. Retainer clip
5. Retaining nut
6. Main shift lever
7. Sleeve
8. Spring
9. Snap ring
10. Pin
11. Tower cover
12. Fill plug
13. High/low shift lever
14. Cover
15. Retaining nut
16. Pin
17. Sleeve
18. Top cover

Fig. 168—Exploded view of shift rails and forks for eight-speed transmission with synchronized third and fourth speeds in the main transmission.

1. Shift rail (High/Low)
2. Shift gate
3. Shift fork
4. Detent springs
   and plungers
5. Retaining screws
6. Shift rail (Rev./1st)
7. Shift gate
8. Shift fork
9. Shift rail (2nd)
10. Shift fork
11. Shift gate
12. Plate
13. Gasket
14. Interlock plunger
    & balls
15. Retainer
16. Shift rail (3rd/4th)
17. Shift gate
18. Shift fork

Fig. 169—Drawing of shift rails, installed. Notice position of notches in plate (12) and location of neutral start switch (N). Plunger of interlock (14) is located in rail (9) and balls are located in retainer (15). Refer to Fig. 168 for legend.

PLANETARY UNIT

Models With Eight-Speed Synchromesh Transmission

175. To remove the planetary unit, first refer to paragraph 158 or paragraph 159 and separate tractor between rear of transmission and front of axle center housing. Remove set screw (5—Fig. 168) from fork (3), then remove the fork and coupler from rear of planetary. Remove the four retaining cap screws and withdraw rear cover (13—Fig. 170), rear thrust ring (6) and planet carrier (7). Pry planetary ring gear (3) with dowels from rear of housing. Remove planetary front cover (2) and shim (1).

To disassemble planet carrier (7), press planet pins (8) forward out of carrier. Remove planet gears (10) with needle rollers (12) and thrust washers (9). Splined sleeve (15) is retained in hub of carrier (7) by retaining ring (16).

When reassembling, be sure to account for all of the needle rollers (12). Each pinion contains two paths of rollers separated by a washer (11) and each roller path contains 27 rollers (12). Use petroleum jelly (not heavy grease) to hold rollers in place while pressing planet pins into carrier.

Be sure that front plate (2) and rear plate (13) are positioned with grooved side toward planet carrier. Slots in front plate and shim (1) must be aligned and toward top when installed on transmission. Use petroleum jelly to hold thrust rings (6) in place. Make certain that tangs of thrust rings engage notches in pinion carrier and that brass side faces away from carrier. Cutaway section of rear plate (13—Fig. 170) should be positioned at the lower, **right** corner of planetary as shown. A lockwasher is used only on mounting screw at corner of cutaway. Tighten mounting screws to 40-47 N•m (30-35 ft.-lbs.) torque. Complete assembly procedure by reversing disassembly.

Fig. 170—Exploded view of high/low range planetary unit used.

1. Front shim
2. Front plate
3. Ring gear
4. Dowel
6. Thrust washer
7. Planet carrier
8. Pinion shaft
9. Thrust washer
10. Pinion
11. Spacer washer
12. Needle rollers (27/path)
13. Rear plate
14. Shift coupler
15. Splined sleeve
16. Retaining ring
17. Transmission mainshaft

PTO INPUT SHAFT AND RETAINER HOUSING

Models With Eight-Speed Synchromesh Transmission

176. To remove the pto input shaft (8—Fig. 171) and retainer (3), first split tractor between engine and transmission as outlined in paragraph 152. Remove clutch release bearing (16), retainer (15), fork

(13) and release shaft (12). Remove brake cross shaft, if so equipped. Remove screws and pull pto input shaft (8) and retainer (3) as a unit from transmission housing.

To disassemble, remove snap ring (6) from rear of retainer, then push input shaft and bearing (5) out of housing toward rear. Bearing can be removed from input shaft after removing snap ring (4). Remove oil seals (1 and 9) and needle bearing (2) only if new parts are to be installed.

Lubricate all seals and bearings with petroleum jelly before assembling. Use special tool (MF 315A or equivalent) to install new needle bearing (2). Install seal (9) with lip toward rear, using special tool (MF 421 or equivalent). Install ball bearing (5) on shaft with shield toward rear, then install snap ring (4). Use special tool (MF255B or equivalent) to install seal (1) with lip toward rear. Use special sleeve (KMF 1004 or equivalent) to protect seal (1), slide input shaft into the retainer housing, then install snap ring (6). Install "O" ring (7) and install housing and shaft assembly over transmission input shaft (11) into housing bore. Coat threads of retaining screws with "Hylomar" or equivalent sealer and tighten to 54-61 N·m (40-45 ft.-lbs.) torque. Remainder of assembly is reverse of disassembly.

PTO LOWER SHAFT FRONT BEARING AND RETAINER

Models With Eight-Speed Synchromesh Transmission

177. The pto lower shaft front bearing (41—Fig. 172) can be removed and installed after separating tractor between the rear of the engine and front of

Fig. 171—Pto input shaft (8) and retainer housing (3) can be removed as a unit from front of transmission housing. Transmission input shaft (11) is removed rearward from transmission housing.

1. Seal
2. Needle bearing
3. Retainer housing
4. Snap ring
5. Ball bearing
6. Snap ring
7. "O" ring
8. Pto input shaft
9. Oil seal
10. Thrust washer
11. Transmission input shaft
12. Clutch lever and shaft
13. Clutch release lever
14. Return spring
15. Release housing
16. Release bearing

Fig. 172—Exploded view of the pto lower shaft and bearings.

36. Cover
37. "O" ring
38. Snap ring
39. Washer
40. Snap ring
41. Ball bearing
42. Retainer
43. Gasket
44. Snap ring
45. Pto lower gear
46. Pto lower shaft

transmission housing. Removal of the lower driven gear (45) and shaft (46) require the additional steps of separating at rear of transmission housing and removal of the transmission top cover and shift rails. The following procedure describes only service to the front bearing and bearing retainer (42) after splitting as outlined in paragraph 152.

Remove clutch release bearing (16—Fig. 171), retainer (15), fork (13) and release shaft (12). Unbolt and remove cover (36—Fig. 172), then remove snap ring (38) and washer (39). Thread a ³⁄₈ inch UNC cap screw, 75 mm (3 inches) long into each of the two threaded holes of retainer (42). Tighten the two jackscrews to push retainer (42) and bearing (41) forward off shaft (46). Bearing can be pressed from retainer after removing snap ring (40). Remove "O" ring (37) and gasket (43), then clean all parts, thoroughly.

Coat gasket (43) and threads of retaining screws lightly with "Hylomar" or equivalent sealer before installing. End of shaft (46) is threaded to accept a puller screw to pull bearing (41) onto shaft until washer (39) and snap ring (38) can be installed. Align holes in retainer (42) with holes in transmission housing, install "O" ring (37) and cover (36). Tighten retaining screws to 54-61 N·m (40-45 ft.-lbs.) torque. Remainder of assembly is reverse of disassembly procedure.

TRANSMISSION INPUT SHAFT

Models With Eight-Speed Synchromesh Transmission

178. To remove the transmission input shaft (11—Fig. 173), first remove the transmission as outlined in paragraph 172. Remove the top cover and shift levers as outlined in paragraph 173, planetary unit as outlined in paragraph 175 and the shift rails and forks as outlined in paragraph 174. Refer to paragraph 177 and remove the pto lower shaft front bearing and retainer. Remove snap ring (44—Fig. 172), pull shaft (46) to rear and allow gear (45) to fall to bottom of housing. Move input shaft (11—Fig. 174) and gear (13) forward, then withdraw gear (13) out top. Transmission input shaft (11) can be removed from same top opening after gear (13) is out.

Reinstall input shaft by reversing removal procedures. Adjust position of shift fork in synchronizer collar as described in paragraph 174.

MAINSHAFT (OUTPUT SHAFT) AND GEARS

Models With Eight-Speed Synchromesh Transmission

179. To remove the mainshaft (19—Fig. 173 and Fig. 174), first remove the transmission as outlined in paragraph 172. Remove the top cover and shift levers as outlined in paragraph 173, planetary unit as outlined in paragraph 175 and the shift rails and forks as outlined in paragraph 174. Refer to paragraph 177 and remove the pto input shaft and retainer, remove the pto lower shaft front bearing and retainer as outlined in paragraph 177, then remove the transmission input shaft as outlined in paragraph 178.

Remove spacer (12—Fig. 174), lift snap ring (59) from groove and move snap ring forward on shaft. Push mainshaft (19) to rear and remove gear (13).

Fig. 173—Cross section of eight-speed transmission with third and fourth speeds of main gear box synchronized.

Remove snap ring (14), then remove bearing (15) and snap ring (16). Continue to move mainshaft toward rear while working snap ring (59) forward on shaft. Withdraw mainshaft toward rear while lifting gears (20, 21 and 24) and synchronizer assembly out top opening. Rear bearing (22) can be removed if necessary after removing snap ring (23).

Synchronizer parts (64 through 67—Fig. 175) should be removed as an assembly and separated only if necessary. Wrap synchronizer assembly with cloth to catch poppet parts (65 and 66) when coupling (64) is removed from hub (67). Clean, inspect and renew any components that are damaged or worn excessively. Place synchronizer cones (72 and 73) and synchronizer rings (63 and 68) on respective gears (21 and 24). Make sure that each ring is seated squarely on taper of cone. Measure clearance between ring and gear at several places around gear, using a feeler gauge. Install new synchronizer ring (63 or 68) if clearance is less than 0.5 mm (0.020 inch). This clearance can be checked with gearbox assembled, but minimum clearance is increased to 0.8 mm (0.030 inch).

When reassembling the synchronizer shift components, align centralizing holes in shift collar (64) and hub (67). Special tools (MF 415 and MS 550) are available to assist holding the three poppets (65 and 66) into hub (67) while installing coupling (64).

Reassemble by reversing removal procedure. Be sure that sleeves (61 and 70) match respective gears (21 and 24) in length. Refer to Fig. 173 for cross section of assembled transmission. Coat gasket (43—

Fig. 172) lightly with "Hylomar" or equivalent sealer before installing. Coat threads of screws attaching cover (36) and screws attaching retainer housing (3—Fig. 171) lightly with "Hylomar" or equivalent sealer before tightening to 54-61 N•m (40-45 ft.-lbs.) torque.

COUNTERSHAFT (LAYSHAFT) AND GEARS

Models With Eight-Speed Synchromesh Transmission

180. The mainshaft must be removed as outlined in paragraph 179 before removing countershaft (31—Fig. 174). Remove snap ring (25) from rear, bump countershaft forward enough to remove snap ring (29) from groove and move snap ring forward to unsplined part of countershaft. Remove snap ring (35), then remove gears (33 and 34) while driving countershaft toward rear. After gears (33 and 34) are lifted from housing, drive countershaft toward front while removing gears (27 and 28).

Reassemble by reversing removal procedure. Snap ring (29) should be around shaft in the unsplined area and front bearing (32) should be in place on shaft before installing shaft in housing. Refer to Fig. 173 for cross section of assembled transmission.

Fig. 174—Exploded view of eight-speed synchronized transmission mainshaft and countershaft.

11. Transmission input shaft	23. Snap ring	61. Sleeve
12. Spacer	24. Gear (3rd) and	62. Bushing
13. Gear (2nd)	synchronizer cone	63. Synchronizer ring
14. Snap ring	25. Snap ring	(Same as 68)
15. Ball bearing	26. Ball bearing	64. Shifting coupler
16. Snap ring	27. Gear (3rd)	65. Pressure blocks (3 used)
17. Spacer	28. Gear (4th)	66. Ball pins & springs (3 used)
18. Roller bearing	29. Snap ring	67. Synchronizer hub
19. Main shaft	30. Roller bearing	68. Synchronizer ring
20. Gear (1st & Reverse)	31. Countershaft &	(Same as 63)
21. Gear (4th) and	1st gear	69. Bushing
synchronizer cone	32. Ball bearing	70. Sleeve
22. Ball bearing	33. Gear (2nd)	71. Washer

34. Constant mesh gear
35. Snap ring
50. Thrust washers
51. Washers
52. Center spacer
53. Needle rollers
54. Gear (Reverse idler)
56. Spacer
57. Idler shaft
58. Retaining clip
59. Snap ring
60. Washer

REVERSE IDLER GEAR AND SHAFT

Models With Eight-Speed Synchromesh Transmission

181. A dummy shaft should be fabricated to hold parts in place while both removing and installing the reverse idler shaft. Dummy shaft should be 25 mm (1 inch) diameter and 55 mm (2$\frac{3}{16}$ inches) long.

To remove the reverse idler, first remove the mainshaft and countershaft assemblies as outlined in paragraphs 179 and 180. Release locking washer, then remove locking screw and clip (58—Fig. 174). Slide dummy shaft in from front, pushing reverse idler shaft (57) out toward rear. Dummy shaft should be used to hold needle rollers and washers in place while both removing and installing the reverse idler. Be sure to remove all 56 of the needle rollers.

Assemble needle rollers, washers and spacer in reverse idler using petroleum jelly and the dummy shaft. Assemble center spacer (52), one path of 28 needle rollers (53), one end washer (51) and one thrust washer (50), around dummy shaft, in one end of reverse idler gear. Assemble the second path of 28

Fig. 175—Exploded view of synchronizer parts for third and fourth main gearbox speeds. Synchronizer cones (72 and 73) are not separated from gears in Fig. 174. Refer to Fig. 174 for legend.

needle rollers, the other end washer (51) and the remaining thrust washer (50). Center the dummy shaft so that it holds both thrust washers in place, then position assembly in housing and install idler shaft (57). Install clip (58), lock plate and retaining screw. Tighten retaining screw and lock by bending tab of lock plate against flat of screw. Complete reassembly by reversing disassembly procedure.

TRANSMISSION
(TWELVE-SPEED SYNCHROMESH)

350, 355, 360 and 399 Models

Models 350, 355, 360 and 399 are available with a transmission providing twelve forward speeds and four reverse speeds. The twelve forward speeds and four reverse speeds are possible by a manually shifted, high/low range, located ahead of the main gearbox, a main gearbox with three forward speeds and one reverse speed, and finally by a two speed planetary unit which compounds the number of speeds already provided. The high/low range and the top two forward gears of the main gearbox are synchronized. Refer to Fig. 176 for shift pattern of twelve-speed synchronized transmission models.

LUBRICATION

Models With Twelve-Speed Synchromesh Transmission

182. The transmission, rear differential and hydraulic system share common fluid which is contained in the transmission and center housing.

Fig. 176—Shift pattern for twelve-speed synchronized transmission. Only third and fourth speeds of the main transmission and high/low (indicated by tortoise and hare/rabbit) are synchronized.

Recommended lubricant is Massey-Ferguson Super 500 10W-30 oil, Massey-Ferguson Permatran or equivalent transmission/hydraulic oil. Capacity is 47.4 L (12.5 gal.) if equipped with spacer or four-wheel-drive transfer box between transmission and rear axle center housing; 43.4 L (11.5 gal.) if not equipped with transfer box or spacer. Tractors are equipped with dipstick (Fig. 177 or Fig. 178) with marks indicating maximum and minimum oil levels. Level should be maintained at maximum level if tractor is operated on hilly terrain or if using implements which require large quantities of oil.

TRACTOR REAR SPLIT

Models With Twelve-Speed Synchromesh Transmission

183. To separate the rear of the transmission from the rear axle center housing (Fig. 179), refer to paragraph 158 if tractor does not have cab or to paragraph 159 if equipped with cab.

TRANSMISSION REMOVAL

Models With Twelve-Speed Synchromesh Transmission

184. To remove the transmission, first separate the rear of the transmission from rear axle center housing as outlined in paragraph 158 if tractor does not have cab or paragraph 159 if equipped with cab. If not equipped with cab, unbolt and support or remove the

Fig. 177—Transmission oil level of tractors without cab is measured by dipstick located in the side cover as shown. Refer to Fig. 178 for models with cab.

Fig. 178—Transmission oil level of tractors with cab is measured by dipstick that points toward rear as shown.

steering and instrument console. On all models, support the engine and transmission housing separately. Remove screws and stud nuts, then separate transmission from engine.

Reassemble transmission to engine by reversing the splitting procedure and observe the following. Install a M12 alignment stud approximately 100 mm (4 inches) long in each side of the transmission housing to assist in aligning engine to transmission. Turn flywheel to align clutch plate splines with transmission and pto input shaft splines. Install retaining screws and nuts after engine and transmission housing flanges are completely together. Tighten retaining screws and nuts to 115 N•m (85 ft.-lbs.) torque.

Refer to paragraph 158 or 159 and reconnect transmission to rear axle center housing. Tighten all screws retaining the transmission housing to spacer housing, transfer case or rear axle center housing to 112 N•m (83 ft.-lbs.) torque, beginning at top center and progressing in a clockwise direction (as viewed from tractor rear) two times around the mating flange. On four-wheel-drive models, coat threads of drive shaft retaining screws with "Loctite 270," then attach drive shaft flanges and tighten screws to 55-75 N•m (40-55 ft.-lbs.) torque. On all models, check and adjust clutch linkage as outlined in paragraph 151.

TOP COVER

Models With Twelve-Speed Synchromesh Transmission

185. The shift levers and the small shift tower can be removed after first removing floor mat and access panels. Installation of the shift levers and tower is easier if transmission is shifted into neutral before removing. Shift tower (11—Fig. 180) is attached to the top cover (18) by five screws.

Fig. 179—View of tractor with cab separated between the four-wheel-drive transfer case and the rear axle center housing. Special splitting stand and track is shown. Support and splitting procedures are similar for models without cab and without four-wheel drive.

The main shift lever (6) can be removed after removing the tower assembly, cover (3), snap ring (9), spring (8) and pin (10). The high/low shift lever (13) and over/under shift lever (19) can be removed after removing cover (14) and pin (16). Refer to Fig. 181 for cross section of shift levers.

The transmission top cover (18—Fig. 180) can be unbolted and lifted from transmission after first removing the shift levers and tower assembly. On models with cab, it is necessary to split rear of transmission from rear axle center housing as outlined in paragraph 159 or remove the cab assembly. Models without cab must have instrument console and other interfering parts removed before transmission top cover. The reason for removing top cover will determine what additional procedures are necessary.

"Loctite 515 Instant Gasket" or equivalent should be used to seal tower to top cover. Tighten retaining screws to 50-70 N·m (37-52 ft.-lbs.) torque.

SHIFTER RAILS AND FORKS

Models With Twelve-Speed Synchromesh Transmission

186. To remove the shifter rails and forks, first refer to paragraph 158 or paragraph 159 and separate tractor between rear of transmission and front of axle center housing. Refer to paragraph 185 and remove the shift lever and tower assembly and transmission top cover. Models without cab must have instrument

and steering console removed before top cover can be unbolted and removed.

On all models, remove the neutral safety start switch (N—Fig. 183). Remove safety wire and set screws (5—Fig. 182) and lift detent springs and plungers (4) from bores in housing. Unbolt and remove interlock parts (12, 13, 14 and 15—Fig. 183). Slide rail out of housing toward rear and remove gate and shift fork through top. Twist rails if necessary while withdrawing. Fork (8—Fig. 182) cannot be removed until transmission input shaft and gears are removed as outlined in paragraph 190.

When assembling, lubricate rails and make sure that rails do not have burrs, especially around the set screw locations, that would prevent easy and smooth installation. Each of the four shift rails is different from the other rails. Tighten screws retaining the interlock retainer (15—Fig. 182) to 40-47 N·m (30-35 ft.-lbs.) torque and the seven set screws (5) to 34-52 N·m (25-38 ft.-lbs.) torque. Install safety wire through set screws (5) after tightening to prevent loosening. Adjust position of shift fork (18) as follows.

Use special tools (MF 414/2 and MF 414/3 or equivalent) to hold shift rails (6 and 16—Fig. 182 and Fig. 183) in neutral detent position. Insert centralizing tool (MF 414/1 or equivalent) into holes in shift forks (8 and 18—Fig. 182) and into matching hole of sliding synchronizing coupling. Tighten each set screws (5S), while turning centralizing tool, to hold the fork in this center position. Screws (5 and 5S) should be tightened to 34-52 N·m (25-38 ft.-lbs.) torque and locked in position with safety wire. Shift

Fig. 180—Exploded view of transmission top cover, tower cover and shift levers typical of models with twelve-speed transmission.

2. Boot	12. Fill plug
3. Cover	13. High/low shift lever
4. Retainer clip	14. Cover
5. Retaining nut	15. Retaining nut
6. Main shift lever	16. Pin
7. Sleeve	17. Sleeve
8. Spring	18. Top cover
9. Snap ring	19. Over/Underdrive
10. Pin	shift lever
11. Tower cover	

all gears, rails and gates to neutral after checking to make sure that all shift gates align. Shift into each gear to make sure that each can be correctly engaged. Install detent plungers and springs (4—Fig. 183). Remainder of assembly is the reverse of disassembly procedure.

Fig. 181—Cross section of shift tower used on tractors with twelve-speed transmission. Refer to Fig. 180 for legend.

PLANETARY UNIT

Models With Twelve-Speed Synchromesh Transmission

187. To remove the planetary unit, first refer to paragraph 158 or paragraph 159 and separate tractor between rear of transmission and front of axle center housing. Remove set screw (5—Fig. 182) from fork (3), then remove the fork and coupler from rear of planetary. Remove the four retaining cap screws and withdraw rear cover (13—Fig. 184 or Fig. 185), rear thrust ring (6) and planet carrier (7). Pry planetary ring gear (3) with dowels from rear of housing. Remove planetary front cover (2) and shim (1).

To disassemble planet carrier (7), first remove snap ring (5—Fig. 185), if used. Press planet pins (8—Fig. 184 or Fig. 185) forward out of carrier. Remove planet gears (10) with needle rollers (12) and thrust washers

Fig. 182—Exploded view of shift rails and forks for twelve-speed transmission with synchronized second and third speeds in the main transmission and synchronized over/under drive.

1. Shift rail (High/Low)
2. Shift gate
3. Shift fork
4. Detent springs
   and plungers
5. Retaining screws
6. Shift rail (Over/
   Underdrive)
7. Shift gate
8. Shift fork
9. Shift rail (1st/Rev.)
10. Shift fork
11. Shift gate
12. Plate
13. Gasket
14. Interlock plunger
    & balls
15. Retainer
16. Shift rail (2nd/3rd)
17. Shift gate
18. Shift fork

(9). Splined sleeve (15) is retained in hub of carrier (7) by retaining ring (16).

When reassembling, be sure to account for all of the needle rollers (12). Each pinion contains two paths of rollers separated by a washer (11). Each roller path contains 27 rollers (12—Fig. 184) on 350, 355 and 360 models. Each roller path contains 16 rollers (12—Fig. 185) on 399 models. Use petroleum jelly (not heavy grease) to hold rollers in place while pressing planet pins into carrier. Gap in snap ring (5—Fig. 185) should be located between planet pins.

Concave side of Belleville washer (18—Fig. 185) should be toward rear. Be sure that front plate (2—Fig. 184 or Fig. 185) and rear plate (13) are positioned with grooved side toward planet carrier. Slots in front plate and shim (1) must be aligned and toward top when installed on transmission. Use petroleum jelly to hold thrust rings (6) in place. Make certain that tangs of thrust rings engage notches in pinion carrier and that brass side faces away from carrier. On 350, 355 and 360 models, cutaway section of rear plate

Fig. 183—Drawing of shift rails, installed. Notice position of notches in plate (12) and location of neutral start switch (N). Plunger of interlock (14) is located in rail (9) and balls are located in retainer (15). Refer to Fig. 182 for legend.

(13—Fig. 184) should be positioned at the lower, **right** corner of planetary as shown. A lockwasher is used only on mounting screw at corner of cutaway. On 399 models, the slanted section of rear plate (13—Fig. 185) should be positioned at lower, **left** corner of planetary. On all models, tighten mounting screws to 40-47 N·m (30-35 ft.-lbs.) torque. Complete assembly procedure by reversing disassembly.

PTO INPUT SHAFT AND RETAINER HOUSING

Models With Twelve-Speed Synchromesh Transmission

188. To remove the pto input shaft (8—Fig. 186) and retainer (3), first split tractor between engine and transmission as outlined in paragraph 152. Remove clutch release bearing (16), retainer (15), fork (13) and release shaft (12). Remove brake cross shaft if so equipped. Remove screws and pull pto input shaft (8) and retainer (3) as a unit from transmission housing.

To disassemble, remove snap ring (6) from rear of retainer, then push input shaft and bearing (5) out of housing toward rear. Bearing can be removed from input shaft after removing snap ring (4). Remove oil seals (1 and 9) and needle bearing (2) only if new parts are to be installed.

Lubricate all seals and bearings with petroleum jelly before assembling. Use special tool (MF 315A or equivalent) to install new needle bearing (2). Install seal (9) with lip toward rear, using special tool (MF 421 or equivalent). Install ball bearing (5) on shaft

with shield toward rear, then install snap ring (4). Use special tool (MF255B or equivalent) to install seal (1) with lip toward rear. Use special sleeve (KMF 1004 or equivalent) to protect seal (1), slide input shaft into the retainer housing, then install snap ring (6). Install "O" ring (7) and install housing and shaft assembly over transmission input shaft (11) into housing bore. Coat threads of retaining screws with "Hylomar" or equivalent sealer and tighten to 54-61

Fig. 184—Exploded view of high/low range planetary unit used on 350, 355 and 360 models. Refer to Fig. 185 for 399 models with twelve-speed synchronized transmission.

1. Front shim	10. Pinion
2. Front plate	11. Spacer washer
3. Ring gear	12. Needle rollers (27/path)
4. Dowel	13. Rear plate
6. Thrust washer	14. Shift coupler
7. Planet carrier	15. Splined sleeve
8. Pinion shaft	16. Retaining ring
9. Thrust washer	17. Transmission mainshaft

Fig. 185—Exploded view of high/low range planetary unit used on 399 models. Refer to Fig. 184 for planetary unit used on other models with twelve-speed synchronized transmission.

1. Front shim	10. Pinion
2. Front plate	11. Spacer washer
3. Ring gear	12. Needle rollers (16/path)
4. Dowel	13. Rear plate
5. Snap ring	14. Shift coupler
6. Thrust washer	15. Splined sleeve
7. Planet carrier	16. Retaining ring
8. Pinion shaft	17. Transmission mainshaft
9. Thrust washer	18. Belleville washer

Fig. 186—Pto input shaft (8) and retainer housing (3) can be removed as a unit from front of transmission housing. Transmission input shaft (11) is removed rearward from transmission housing.

1. Seal
2. Needle bearing
3. Retainer housing
4. Snap ring
5. Ball bearing
6. Snap ring
7. "O" ring
8. Pto input shaft
9. Oil seal
10. Thrust washer
11. Transmission input shaft
12. Clutch lever and shaft
13. Clutch release lever
14. Return spring
15. Release housing
16. Release bearing

N·m (40-45 ft.-lbs.) torque. Remainder of assembly is reverse of disassembly.

PTO LOWER SHAFT FRONT BEARING AND RETAINER

Models With Twelve-Speed Synchromesh Transmission

189. The pto lower shaft front bearing (41—Fig. 187) can be removed and installed after separating tractor between the rear of the engine and front of transmission housing. Removal of the lower driven gear (45) and shaft (46) require the additional steps of separating at rear of transmission housing and removal of the transmission top cover and shift rails. The following procedure describes only service to the front bearing and bearing retainer (42) after splitting as outlined in paragraph 152.

Remove clutch release bearing (16—Fig. 186), retainer (15), fork (13) and release shaft (12). Unbolt and remove cover (36—Fig. 187), then remove snap ring (38) and washer (39). Thread a ⅜ inch UNC cap screw, 75 mm (3 inches) long into each of the two threaded holes of retainer (42). Tighten the two jackscrews to push retainer (42) and bearing (41) forward off shaft (46). Bearing can be pressed from retainer after removing snap ring (40). Remove "O" ring (37) and gasket (43), then clean all parts, thoroughly.

Coat gasket (43) and threads of retaining screws lightly with "Hylomar" or equivalent sealer before installing. End of shaft (46) is threaded to accept a puller screw to pull bearing (41) onto shaft until washer (39) and snap ring (38) can be installed. Align holes in retainer (42) with holes in transmission housing, install "O" ring (37) and cover (36). Tighten retaining screws to 54-61 N·m (40-45 ft.-lbs.) torque.

Remainder of assembly is reverse of disassembly procedure.

TRANSMISSION INPUT SHAFT

Models With Twelve-Speed Synchromesh Transmission

190. To remove the transmission input shaft (11—Fig. 188 and Fig. 189), first remove the transmission as outlined in paragraph 184. Remove the top cover and shift levers as outlined in paragraph 185, planetary unit as outlined in paragraph 187 and the shift rails and forks as outlined in paragraph 186. Refer to paragraph 188 and remove the pto input shaft and retainer. Slide input shaft (11—Fig. 189) forward, then withdraw gears (81 and 82) out top. Over/Under shift fork (8—Fig. 182) can be removed after removing input gears.

Reinstall input shaft by reversing removal procedures. Refer to paragraph 186 for adjusting neutral position of forks on shift shafts.

MAINSHAFT (OUTPUT SHAFT) AND GEARS

Models With Twelve-Speed Synchromesh Transmission

191. To remove the mainshaft (19—Fig. 188 and Fig. 189), first remove the transmission as outlined in paragraph 184. Remove the top cover and shift levers as outlined in paragraph 185, planetary unit as outlined in paragraph 187 and the shift rails and forks as outlined in paragraph 186. Refer to paragraph 188 and remove the pto input shaft and retainer, then remove the transmission input shaft as outlined in paragraph 190.

Fig. 187—Exploded view of the pto lower shaft and bearings.

36. Cover
37. "O" ring
38. Snap ring
39. Washer
40. Snap ring
41. Ball bearing
42. Retainer
43. Gasket
44. Snap ring
45. Pto lower gear
46. Pto lower shaft

Fig. 188—Cross section of twelve-speed transmission.

Remove spacer (83—Fig. 189), lift snap ring (14) from groove, push mainshaft (19) to rear and remove gear (13). Continue to move mainshaft toward rear while lifting synchronizer assembly (91 and 92), bearing (93), gears (20, 21 and 24) and synchronizer assembly (63 through 68) out top opening. Rear bearing (22) can be removed if necessary after removing snap ring (23).

Synchronizer parts (64 through 67—Fig. 190) should be removed as an assembly and separated only if necessary. Wrap synchronizer assembly with a cloth to catch poppet parts (65 and 66) when coupling (64) is removed from hub (67). Clean, inspect and renew any components that are damaged or worn excessively. Place synchronizer cones (72 and 73—Fig. 189) and synchronizer rings (63 and 68) on respective gears (21 and 24). Make sure that each ring is seated squarely on taper of cone. Measure clearance between ring and gear at several places around gear, using a feeler gauge. Install new synchronizer ring (63 or 68) if clearance is less than 0.5 mm (0.020 inch). This clearance can be checked with gearbox assembled, but minimum clearance is increased to 0.8 mm (0.030 inch).

When reassembling the synchronizer shift components, align centralizing holes in shift collar (64) and hub (67). Special tools (MF 415 and MS 550) are available to assist holding the three poppets (65 and 66) into hub (67) while installing coupling (64).

Make sure that Low (Tortoise) and High (Hare) shift fork is in place in shift collar before installing the mainshaft.

Reassemble mainshaft (19—Fig. 189), snap rings (14 and 23), bearings (15 and 22) and all parts between snap rings (14 and 23), outside of gearbox housing. Be sure that sleeves (61 and 70) match respective gears (21 and 24) in length. Refer to Fig. 191 for cross section of assembled mainshaft. Measure clearance (G) between front snap ring (14) and inner race of bearing (15). Select spacer (17—Fig. 189 and Fig. 191) of proper thickness to limit gap (G) to 0.080-0.300 mm (0.003-0.012 inch). Spacer (17) is available in four thicknesses of 4.14-4.19, 4.39-4.44, 4.62-4.67 and 4.85-4.90 mm (0.163-0.165, 0.173-0.175, 0.182-0.184 and 0.191-0.193 inch). Reassemble mainshaft and components inside housing using the preselected spacer (17). Shorter sleeve (70—Fig. 189) is in 2nd gear (24) and longer sleeve (61) is in 3rd and reverse gears (20 and 72). Flat side of spacer (17) should be toward gear (13).

Remainder of assembly is reverse of disassembly procedure. Coat threads of screws attaching retainer housing (3—Fig. 186) lightly with "Hylomar" or equivalent sealer before tightening to 54-61 N·m (40-45 ft.-lbs.) torque.

COUNTERSHAFT (LAYSHAFT) AND GEARS

Models With Twelve-Speed Synchromesh Transmission

192. The mainshaft must be removed as outlined in paragraph 191 and pto lower shaft should be re-moved as outlined in paragraph 189, before removing countershaft (31—Fig. 189). Remove snap ring (25) from rear, bump countershaft forward enough to remove snap ring (29) from groove and move snap ring forward to unsplined part of countershaft. Remove snap ring (35) and thrust washer (84). Slide gear (34) forward, remove snap ring (86) from groove, then slide snap ring forward as far as possible. Slide shaft (31) toward rear as far as possible and remove gear

Fig. 189—Exploded view of twelve-speed transmission. Synchronizer (63 through 68) is made up of same components as synchronizer (75 through 80).

11. Transmission input shaft	28. Gear (3rd)	62. Bushing	78. Ball pins & springs (3 used)
12. Spacer	29. Snap ring	63. Synchronizer ring (Same as 68)	79. Synchronizer hub
13. Gear (1st)	30. Roller bearing	64. Shifting coupler	80. Synchronizer ring (Same as 75)
14. Snap ring	31. Countershaft & 1st gear	65. Pressure blocks (3 used)	81. Gear (Low/Tortoise)
15. Bearing	32. Ball bearing	66. Ball pins & springs (3 used)	82. Gear (High/Hare)
16. Snap ring	33. Gear (Low/Tortoise)	67. Synchronizer hub	83. Spacer
17. Spacer (variable thicknesses)	34. Gear (High/Hare)	68. Synchronizer ring (Same as 63)	84. Thrust washer
18. Roller bearing	35. Snap ring	69. Bushing	85. Bushing
19. Mainshaft	50. Thrust washer	70. Sleeve (short)	86. Snap ring
20. Gear (Reverse)	51. Washers	71. Washer	87. Bushing
21. Gear (3rd)	52. Center spacer	72. Synchronizer cone	88. Thrust washer
22. Ball bearing	53. Needle rollers	73. Synchronizer cone	89. Sleeve (medium)
23. Snap ring	54. Gear (Reverse idler)	75. Synchronizer ring (Same as 80)	90. Bushing
24. Gear (2nd)	55. Friction disc	76. Shifting coupler	91. Shifting coupler
25. Snap ring	56. Spacer	77. Pressure blocks (3 used)	92. Hub
26. Ball bearing	57. Idler shaft		93. Thrust bearing
27. Gear (2nd)	58. Retaining clip		94. Bushing
	60. Washer		
	61. Sleeve (longer)		

Fig. 190—Exploded view of synchronizer parts for 2nd and 3rd main gearbox speeds and Low/Tortoise and High/Hare speeds. Refer to Fig. 189 for legend.

Fig. 191—The mainshaft should be assembled outside of transmission housing to determine correct thickness of thrust washer (17) as determined by gap (G).

(34). Remove snap ring (86) and synchronizer ring (75), then remove synchronizer assembly (76 through 79) as a unit. Continue sliding shaft (31) toward rear and remove synchronizer ring (80), gear (33) and washer (88). Slide shaft (31) toward front and remove gears (27 and 28).

Synchronizer parts (76 through 79—Fig. 190) should be removed as an assembly and separated only if necessary. Wrap synchronizer assembly with a cloth to catch poppet parts (77 and 78) when coupling (76) is removed from hub (79). Clean, inspect and renew any components that are damaged or worn excessively. Place synchronizer rings (75 and 80—Fig. 189) on respective gears (33 and 34). Make sure that each ring is seated squarely on taper of cone. Measure clearance between ring and gear at several places around gear, using a feeler gauge. Install new synchronizer ring if clearance is less than 0.5 mm (0.020 inch). This clearance can be checked with gearbox assembled, but minimum clearance is increased to 0.8 mm (0.030 inch).

When reassembling the synchronizer shift components, align centralizing holes in shift collar (76) and hub (79). Special tools (MF 415 and MS 550) are available to assist holding the three poppets (77 and 78) into hub (79) while installing coupling (76).

Reassemble by reversing removal procedure. Always install new snap rings when reassembling. Position snap ring (29) around shaft in the unsplined area and install front bearing (32) on shaft before installing shaft in housing. Refer to Fig. 188 for cross section of assembled transmission. Refer to paragraph 191 for assembly of mainshaft and selection of thrust washer (17—Fig. 189 and Fig. 191).

Coat gasket (43—Fig. 187) lightly with "Hylomar" or equivalent sealer before installing. Coat threads of screws attaching cover (36) and screws attaching retainer housing (3—Fig. 186) lightly with "Hylomar" or equivalent sealer before tightening to 54-61 N·m (40-45 ft.-lbs.) torque.

REVERSE IDLER GEAR AND SHAFT

Models With Twelve-Speed Synchromesh Transmission

193. A dummy shaft should be fabricated to hold parts in place while both removing and installing the reverse idler shaft. Dummy shaft should be 25 mm (1 inch) diameter and 55 mm (2³⁄₁₆ inches) long.

To remove the reverse idler, first remove the mainshaft and countershaft assemblies as outlined in paragraphs 191 and 192. Release locking washer, then remove locking screw and clip (58—Fig. 189). Slide dummy shaft in from front, pushing reverse idler shaft (57) out toward rear. Dummy shaft should be used to hold needle rollers and washers in place while both removing and installing the reverse idler. Be sure to remove all 56 of the needle rollers.

Assemble needle rollers, washers and spacer in reverse idler using petroleum jelly and the dummy shaft. Assemble center spacer (52), one path of 28 needle rollers (53), one end washer (51) and one thrust washer (50), around dummy shaft, in one end of reverse idler gear. Assemble the second path of 28 needle rollers, the other end washer (51) and the remaining thrust washer (50). Center the dummy shaft so that it holds both thrust washers in place, then position assembly in housing and install idler shaft (57). Install clip (58), lock plate and retaining screw. Tighten retaining screw and lock by bending tab of lock plate against flat of screw. Complete reassembly by reversing disassembly procedure.

TRANSMISSION
(SIX-SPEED SHUTTLE)

360 Models

Six forward speeds are possible by using a standard gearbox with three forward speeds compounded by a two speed planetary unit. An equal number of reverse speeds is made possible by fitting a forward/reverse unit in front of the gearbox. Refer to Fig. 192 for shift pattern of six-speed shuttle transmission available in 360 model tractors.

LUBRICATION

Models With Six-Speed Manual Shuttle Transmission

194. The transmission, rear differential and hydraulic system share common fluid which is contained in the transmission and center housing. Recommended lubricant is Massey-Ferguson Super 500 multi-use 10W-30 oil, Massey-Ferguson Permatran or equivalent transmission/hydraulic oil. Capacity is 47.4 L (12.5 gal.) if equipped with spacer or four-wheel-drive transfer box between transmission and rear axle center housing; 43.4 L (11.5 gal.) if not equipped with transfer box or spacer. Tractors are equipped with dipstick (Fig. 193 or Fig. 194) with marks indicating maximum and minimum oil levels. Level should be maintained at maximum level if tractor is operated on hilly terrain or if using implements which require large quantities of oil.

TRACTOR REAR SPLIT

Models With Six-Speed Manual Shuttle Transmission

195. To separate the rear of the transmission from the rear axle center housing (Fig. 195), refer to paragraph 158 if tractor does not have cab or to paragraph 159 if equipped with cab.

TRANSMISSION REMOVAL

Models With Six-Speed Manual Shuttle Transmission

196. To remove the transmission (Fig. 195), first separate the rear of the transmission from rear axle center housing as outlined in paragraph 158 if tractor does not have cab or paragraph 159 if equipped with cab. If not equipped with cab, unbolt and support or

Fig. 193—Transmission oil level of tractors without cab is measured by dipstick located in the side cover as shown. Refer to Fig. 194 for models with cab.

Fig. 192—Shift pattern for 6 speed shuttle transmission.

Fig. 194—Transmission oil level of tractors with cab is measured by dipstick that points toward rear as shown.

remove the steering and instrument console. On all models, support the engine and transmission housing separately. Remove screws and stud nuts, then separate transmission from engine.

Reassemble transmission to engine by reversing the splitting procedure and observe the following. Install a M12 alignment stud approximately 100 mm (4 inches) long in each side of the transmission housing to assist in aligning engine to transmission. Turn flywheel to align clutch plate splines with transmission and pto input shaft splines. Install retaining screws and nuts after engine and transmission housing flanges are completely together. Tighten retaining screws and nuts to 115 N•m (85 ft.-lbs.) torque.

Refer to paragraph 158 or 159 and reconnect transmission to rear axle center housing. Tighten all screws retaining the transmission housing to spacer housing, transfer case or rear axle center housing to 112 N•m (83 ft.-lbs.) torque, beginning at top center and progressing in a clockwise direction (as viewed from tractor rear) two times around the mating flange. On four-wheel-drive models, coat threads of drive shaft retaining screws with "Loctite 270," then attach drive shaft flanges and tighten screws to 55-75

N•m (40-55 ft.-lbs.) torque. On all models, check and adjust clutch linkage as outlined in paragraph 151.

TOP COVER

Models With Six-Speed Manual Shuttle Transmission

197. The shift levers and the small shift tower can be removed after first removing floor mat and access panels. Installation of the shift levers and tower is easier if transmission is shifted into neutral before removing. Shift tower (11—Fig. 196) is attached to the top cover (18) by five screws.

The main shift lever (6) can be removed after removing the tower assembly, cover (3), snap ring (9), spring (8) and pin (10). The high/low shift lever (13) and over/under shift lever (19) can be removed after removing cover (14) and pin (16). Refer to Fig. 197 for cross section of shift levers.

The transmission top cover (18—Fig. 196) can be unbolted and lifted from transmission after first removing the shift levers and tower assembly. On models with cab, it is necessary to split rear of transmission from rear axle center housing as outlined in paragraph 159 or remove the cab assembly. Models without cab must have instrument console and other interfering parts removed before transmission top cover can be removed. The reason for removing top cover will determine what additional procedures are necessary.

"Loctite 515 Instant Gasket" or equivalent should be used to seal tower to top cover. Tighten retaining screws to 50-70 N•m (37-52 ft.-lbs.) torque. Typical shift linkage for shuttle (forward and reverse shift) is shown in Fig. 198.

SHIFTER RAILS AND FORKS

Models With Six-Speed Manual Shuttle Transmission

198. To remove the shift rails and forks, first separate the rear of transmission from the rear axle center housing as outlined in paragraph 158 if tractor does not have cab or paragraph 159 if equipped with cab. Remove transmission top cover as outlined in paragraph 197.

Carefully remove the detent assemblies (4—Fig. 200) and neutral start switch (N). Remove safety wire from screws (5—Fig. 199), then loosen screws from all shift forks and gates. Remove the planetary shift fork (3), then unbolt and remove the shift rail interlock (15). Pull the shuttle and planetary shift rail (1) out rear of transmission and withdraw planetary gate (2) and shuttle fork (8) from top. Slide the remaining shift rails out of shift forks, gates and transmission housing. Twist rails if necessary while withdrawing.

Fig. 195—View of tractor with standard cab separated between the transmission housing and the rear axle center housing. Special splitting stand and track is shown. Support and splitting procedures are similar for other models.

Fig. 196—Exploded view of transmission top cover, tower cover and shift levers typical of models with 6 speed shuttle transmission.

   3. Cover
   4. Retainer clip
   5. Retaining nut
   6. Main shift lever
   7. Sleeve
   8. Spring
   9. Snap ring
  10. Pin
  11. Tower cover
  12. Fill plug
  13. Forward/Reverse shift lever
  14. Cover
  15. Retaining nut
  16. Pin
  17. Sleeve
  18. Top cover
  19. High/Low shift lever

Fig. 197—Cross section of shift tower used on tractors with six-speed shuttle transmission. Refer to Fig. 196 for legend.

Fig. 198—Forward/Reverse shift lever (13—Fig. 196) is operated by linkage shown for some models.

Fig. 199—Exploded view of shift rails and forks for six-speed shuttle transmission.

1. Shift rail (High/Low)	10. Shift fork (1st)
2. Shift gate	11. Shift gate
3. Shift fork	14. Interlock plunger
4. Detent springs	15. Retainer
and plungers	16. Shift rail (2nd/3rd)
5. Retaining screws	17. Shift gate
8. Shift fork (Forward/Reverse)	18. Shift fork (2nd/3rd)

When assembling, lubricate rails and make sure that rails do not have burrs, especially around the set screw locations, that would prevent easy and smooth installation. Each of the three shift rails is different from the other rails. It may be necessary to move the shuttle coupling forward so that fork may be inserted into groove. Tighten screws retaining the interlock retainer (15—Fig. 200) to 40-47 N•m (30-35 ft.-lbs.) torque and the set screws (5—Fig. 199) to 34-52 N•m (25-38 ft.-lbs.) torque. Install safety wire through set screws (5) after tightening to prevent loosening.

PLANETARY UNIT

Models With Six-Speed Manual Shuttle Transmission

199. To remove the planetary unit, first refer to paragraph 158 or paragraph 159 and separate tractor between rear of transmission and front of axle center housing. Remove set screw from fork (3—Fig. 200), then remove the fork and coupler from rear of planetary. Remove the four retaining cap screws and with-

Fig. 200—Drawing of shift rails, installed. Notice position of neutral start switch (N). Interlock plunger (14) is located retainer (15). Refer to Fig. 199 for legend.

Fig. 201—Exploded view of planetary assembly used on 360 models with six-speed shuttle transmission.

1. Front shim	10. Pinion
2. Front plate	11. Spacer washer
3. Ring gear	12. Needle rollers (16/path)
4. Dowel	13. Rear plate
5. Snap ring	14. Shift coupler
6. Thrust washer	15. Splined sleeve
7. Planet carrier	16. Retaining ring
8. Pinion shaft	17. Transmission mainshaft
9. Thrust washer	18. Belleville washer

draw rear cover (13—Fig. 201), rear thrust ring (6) and planet carrier (7). Pry planetary ring gear (3) with dowels from rear of housing. Remove planetary front cover (2) and shim (1).

To disassemble planet carrier (7), first remove snap ring (5). Press planet pins (8) forward out of carrier. Remove planet gears (10) with needle rollers (12) and thrust washers (9). Splined sleeve (15) is retained in hub of carrier (7) by retaining ring (16).

When reassembling, be sure to account for all of the needle rollers (12). Each pinion contains two paths of rollers separated by a washer (11). Each roller path contains 16 rollers (12). Use petroleum jelly (not heavy grease) to hold rollers in place while pressing planet pins into carrier. Gap in snap ring (5) should be located between planet pins.

Be sure that front plate (2) and rear plate (13) are positioned with grooved side toward planet carrier. Slots in front plate and shim (1) must be aligned and toward top when installed on transmission. Use petroleum jelly to hold thrust rings (6) in place. Make certain that tangs of thrust rings engage notches in pinion carrier and that brass side faces away from carrier. The slanted section of rear plate (13) should be positioned at lower, **left** corner of planetary. Tighten mounting screws to 40-47 N·m (30-35 ft.-lbs.) torque. Complete assembly procedure by reversing disassembly.

PTO INPUT SHAFT AND RETAINER HOUSING

Models With Six-Speed Manual Shuttle Transmission

200. To remove the pto input shaft (8—Fig. 202) and retainer (3), first split tractor between engine and transmission as outlined in paragraph 152. Remove clutch release bearing (16), retainer (15), fork (13) and release shaft (12). Remove brake cross shaft, if so equipped. Remove screws and pull pto input shaft (8) and retainer (3) as a unit from transmission housing.

To disassemble, remove snap ring (6) from rear of retainer, then push input shaft and bearing (5) out of housing toward rear. Bearing can be removed from input shaft after removing snap ring (4). Remove oil seals (1 and 9) and needle bearing (2) only if new parts are to be installed.

Lubricate all seals and bearings with petroleum jelly before assembling. Use special tool (MF 315A or equivalent) to install new needle bearing (2). Install seal (9) with lip toward rear, using special tool (MF 421 or equivalent). Install ball bearing (5) on shaft with shield toward rear, then install snap ring (4). Use special tool (MF255B or equivalent) to install seal (1) with lip toward rear. Use special sleeve (KMF

Fig. 202—Pto input shaft (8) and retainer housing (3) can be removed as a unit from front of transmission housing.

1. Seal
2. Needle bearing
3. Retainer housing
4. Snap ring
5. Ball bearing
6. Snap ring
7. "O" ring
8. Pto input shaft
9. Oil seal
10. Thrust washer
11. Transmission input shaft
12. Clutch lever and shaft
13. Clutch release lever
14. Return spring
15. Release housing

1004 or equivalent) to protect seal (1), slide input shaft into the retainer housing, then install snap ring (6). Install "O" ring (7) and install housing and shaft assembly over transmission input shaft (11) into housing bore. Coat threads of retaining screws with "Hylomar" or equivalent sealer and tighten to 54-61 N•m (40-45 ft.-lbs.) torque. Remainder of assembly is reverse of disassembly.

PTO LOWER SHAFT FRONT BEARING AND RETAINER

Models With Six-Speed Manual Shuttle Transmission

201. The pto lower shaft front bearing (41—Fig. 203) can be removed and installed after separating tractor between the rear of the engine and front of transmission housing. Removal of the lower driven gear (45) and shaft (46) require the additional steps of separating at rear of transmission housing and removal of the transmission top cover and shift rails. The following procedure describes only service to the front bearing and bearing retainer (42) after splitting as outlined in paragraph 152.

Remove clutch release bearing (16—Fig. 202), retainer (15), fork (13) and release shaft (12). Unbolt and remove cover (36—Fig. 203), then remove snap ring (38) and washer (39). Thread a ⅜ inch UNC cap screw, 75 mm (3 inches) long into each of the two threaded holes of retainer (42). Tighten the two jackscrews to push retainer (42) and bearing (41) forward off shaft (46). Bearing can be pressed from retainer after removing snap ring (40). Remove "O" ring (37) and gasket (43), then clean all parts, thoroughly.

Coat gasket (43) and threads of retaining screws lightly with "Hylomar" or equivalent sealer before installing. End of shaft (46) is threaded to accept a puller screw to pull bearing (41) onto shaft until washer (39) and snap ring (38) can be installed. Align holes in retainer (42) with holes in transmission housing, install "O" ring (37) and cover (36). Tighten retaining screws to 54-61 N•m (40-45 ft.-lbs.) torque. Remainder of assembly is reverse of disassembly procedure.

TRANSMISSION INPUT SHAFT AND SHUTTLE GEARS

Models With Six-Speed Manual Shuttle Transmission

202. To remove the transmission input shaft (11—Fig. 203 and Fig. 204), first remove the transmission as outlined in paragraph 196. Remove the top cover and shift levers as outlined in paragraph 197, planetary unit as outlined in paragraph 199 and the shift rails and forks as outlined in paragraph 198. Refer to paragraph 201 and remove the pto lower shaft front bearing and retainer. Remove snap ring (44—Fig. 203), pull shaft (46) to rear and allow gear (45) to fall to bottom of housing. Withdraw input shaft (11), gear (82), coupling (76) and gear (81) forward. Be careful not to drop any parts, especially detents and bearing parts, while removing.

Reinstall input shaft by reversing removal procedures. Assemble bearings (13, 14 and 15) inside gears (81 and 82) using petroleum jelly. Each path of rollers (14) contains 60 rollers. Also use petroleum jelly when installing bearing (17 and 18). Thrust washer (12) is available in thicknesses of 1.50-1.57, 1.75-1.82, 2.00-2.08, 2.26-2.33, 2.51-2.59 and 2.76-2.84 mm (0.59-0.062, 0.069-0.072, 0.079-0.082, 0.089-0.092, 0.099-0.102 and 0.109-0.112 inch). Select thickness of thrust washer (12) which will provide transmission input shaft with 0.25-0.50 mm (0.010-0.020 inch) end play. Coat gasket and threads of screws attaching retainer (3) with "Hylomar" or equivalent sealer and tighten screws to 54-61 N·m (40-45 ft.-lbs.) torque.

MAINSHAFT (OUTPUT SHAFT) AND GEARS

Models With Six-Speed Manual Shuttle Transmission

203. To remove the mainshaft (19—Fig. 203) and associated gears, first remove the transmission as

Fig. 203—Exploded view of six-speed shuttle transmission. Thrust washer (12) is available in several thicknesses, to adjust end play. Bearing parts (13, 14 and 15) are also in gear (81).

3. Retainer housing	18. Bearing rollers (23)	32. Ball bearing	46. Pto lower shaft
4. Snap ring	19. Mainshaft	34. Constant mesh gear	50. Thrust washer
5. Ball bearing	20. Gear (2nd/3rd)	35. Snap ring	51. Washers
6. Snap ring	21. Gear (1st)	36. Cover	52. Center spacer
8. Pto input shaft	22. Ball bearing	37. "O" ring	53. Needle rollers (23/path)
9. Seal	23. Snap rings	38. Snap ring	54. Reverse idler gear
10. Needle bearing	24. Gear (1st)	39. Washer	56. "O" ring
11. Transmission input shaft	25. Snap ring	40. Snap ring	57. Idler shaft
12. Thrust washer	26. Ball bearing	41. Ball bearing	76. Shifting coupler
13. Rings	27. Gear (2nd)	42. Retainer	77. Detent pins (2 used)
14. Needle rollers (60/path)	28. Gear (3rd)	43. Gasket	78. Spring
15. Center spacer	30. Roller bearing	44. Snap ring	81. Forward (shuttle) gear
16. Ball bearing	31. Countershaft	45. Pto lower gear	82. Reverse (shuttle) gear
17. Rings			83. Spacer

Fig. 204—Cross section of six-speed shuttle transmission.

outlined in paragraph 196, the top cover and shift levers as outlined in paragraph 197, planetary unit as outlined in paragraph 199, the shift rails and forks as outlined in paragraph 198, the pto lower shaft front bearing and retainer as outlined in paragraph 201 and the transmission input shaft as outlined in paragraph 202.

Tap mainshaft (19) to the rear, to force bearing (16) from shaft. Continue to slide shaft to rear while removing gears (20 and 21). Rear bearing (22) should be pressed forward off shaft after removing the front snap ring (23), if renewal is required.

To reinstall shaft, reverse removal procedure. Large external snap ring on bearing (22) should be toward rear. Refer to Fig. 204 for cross sectional view.

COUNTERSHAFT (LAYSHAFT) AND GEARS

Models With Six-Speed Manual Shuttle Transmission

204. Refer to paragraph 203 and remove mainshaft. Be sure that pto shaft (46—Fig. 203) and pto constant mesh gear (45) are removed, then remove snap ring (35) from front of countershaft (31). Tap countershaft toward rear, then remove constant mesh gear (34) from front of shaft and snap ring (25) from rear. Push shaft forward and lift gears (24, 27 and 28) from housing. Front bearing (32) can be pressed from shaft and rear bearing (26) can be removed from housing.

When reassembling, reverse the disassembly procedure. Position gear (28) with beveled edge of teeth and hub toward rear. Position gear (27) with beveled edge of teeth and longer hub forward. Install rear (1st) gear (24) with beveled edge of teeth and hub toward rear.

REVERSE IDLER GEAR AND SHAFT

Models With Six-Speed Manual Shuttle Transmission

205. A dummy shaft can be fabricated to hold parts in place while both removing and installing the reverse idler shaft. Dummy shaft should be 25 mm (1 inch) diameter and approximately 100 mm (4 inches) long.

To remove reverse idler gear (54—Fig. 203), shaft (57) and related parts, first remove the transmission as outlined in paragraph 196. Remove the top cover and shift levers as outlined in paragraph 197, planetary unit as outlined in paragraph 199 and the shift rails and forks as outlined in paragraph 198. Refer to paragraph 201 and remove the pto lower shaft front bearing and retainer, then refer to paragraph 202 and remove the transmission input shaft. Remove the spacer (83) from front of mainshaft.

Insert dummy shaft into idler gear (54) from rear and push idler shaft (57) out of housing and gear. Remove the idler gear, bearings and dummy shaft through top opening, being careful not to drop bearings or other parts into transmission housing. Each bearing path (53) contains 23 rollers.

Assemble needle rollers, washers and spacer in reverse idler using petroleum jelly and the dummy shaft. Assemble center spacer (52), one path of 23 needle rollers (53), one end washer (51) and one thrust washer (50), around dummy shaft, in one end of reverse idler gear. Assemble the second path of 23

needle rollers, the other end washer (51) and the remaining thrust washer (50). Center the dummy shaft so that it holds both thrust washers in place then position assembly in housing and install idler shaft (57), using new "O" ring (56). Complete reassembly by reversing disassembly procedure.

TRANSMISSION
(EIGHT-SPEED SHUTTLE)

340, 350, 355 and 360 Models

Eight forward speeds are possible by using a standard gearbox with four forward speeds compounded by a two speed planetary unit. An equal number of reverse speeds is made possible by fitting a forward/reverse unit in front of the gearbox. Refer to Fig. 205 for shift pattern of eight-speed shuttle transmission available in 340, 350, 355 and 360 models tractors.

LUBRICATION

Models With Eight-Speed Shuttle Transmission

206. The transmission, rear differential and hydraulic system share common fluid which is contained in the transmission and center housing. Recommended lubricant is Massey-Ferguson Super 500 multi-use 10W-30 oil, Massey-Ferguson Permatran or equivalent transmission/hydraulic oil. Capacity is 47.4 L (12.5 gal.) if equipped with spacer or four-wheel-drive transfer box between transmission and rear axle center housing; 43.4 L (11.5 gal.) if not equipped with transfer box or spacer. Tractors are equipped with dipstick (Fig. 206 or Fig. 207) with marks indicating maximum and minimum oil levels. Level should be maintained at maximum level if tractor is operated on hilly terrain or if using implements which require large quantities of oil.

TRACTOR REAR SPLIT

Models With Eight-Speed Shuttle Transmission

207. To separate the rear of the transmission from the rear axle center housing (Fig. 208), refer to paragraph 158 if tractor does not have cab or to paragraph 159 if equipped with cab.

TRANSMISSION REMOVAL

Models With Eight-Speed Shuttle Transmission

208. To remove the transmission, first separate the rear of the transmission from rear axle center housing as outlined in paragraph 158 if tractor does not have cab or paragraph 159 if equipped with cab. If not equipped with cab, unbolt and support or remove the steering and instrument console. On all models, support the engine and transmission housing separately. Remove screws and stud nuts, then separate transmission from engine.

Reassemble transmission to engine by reversing the splitting procedure and observe the following. Install a M12 alignment stud approximately 100 mm (4 inches) long in each side of the transmission housing to assist in aligning engine to transmission. Turn flywheel to align clutch plate splines with transmission and pto input shaft splines. Install retaining screws and nuts after engine and transmission housing flanges are completely together. Tighten retaining screws and nuts to 115 N·m (85 ft.-lbs.) torque.

Refer to paragraph 158 or 159 and reconnect transmission to rear axle center housing. Tighten all screws retaining the transmission housing to spacer housing, transfer case or rear axle center housing to 112 N·m (83 ft.-lbs.) torque, beginning at top center and progressing in a clockwise direction (as viewed from tractor rear) two times around the mating flange. On four-wheel-drive models, coat threads of drive shaft retaining screws with "Loctite 270," then attach drive shaft flanges and tighten screws to 55-75

Fig. 205—Shift pattern for eight-speed shuttle transmission.

Fig. 207—Transmission oil level of tractors with cab is measured by dipstick that points toward rear as shown.

Fig. 206—Transmission oil level of tractors without cab is measured by dipstick located in the side cover as shown. Refer to Fig. 207 for models with cab.

N•m (40-55 ft.-lbs.) torque. On all models, check and adjust clutch linkage as outlined in paragraph 151.

TOP COVER

Models With Eight-Speed Shuttle Transmission

209. The shift levers and the small shift tower can be removed after first removing floor mat and access panels. Disconnect Forward and Reverse shift linkage from shift lever. Installation of the shift levers and tower is easier if transmission is shifted into neutral before removing. Shift tower (11—Fig. 209) is attached to the top cover (18) by five screws.

The main shift lever (6) can be removed after removing the tower assembly, cover (3), snap ring (9), spring (8) and pin (10). The high/low shift lever (13) and forward/reverse shift lever (19) can be removed after removing cover (14) and pin (16).

Refer to Fig. 210 for forward/reverse shuttle shift linkage. Be sure that electrical wiring does not touch any part of shift linkage. Lubricate linkage with general purpose grease when assembling. Adjust lengths of rods attached to the relay lever as shown in Fig. 211. Install shims between relay lever pivot and retaining snap ring to remove any end play that would result in lost motion.

The transmission top cover (18—Fig. 209) can be unbolted and lifted from transmission after first removing the shift levers and tower assembly. On models with cab, it is necessary to split rear of transmission from rear axle center housing as outlined in paragraph 159 or remove the cab assembly. Models without cab must have instrument console and other interfering parts removed before transmis-

Fig. 208—View of tractor with standard cab separated between the transmission housing and the rear axle center housing. Special splitting stand and track is shown. Support and splitting procedures are similar for other models.

Fig. 209—Exploded view of transmission top cover, tower cover and shift levers typical of models with eight-speed shuttle transmission.

  3. Cover
  4. Retainer clip
  5. Retaining nut
  6. Main shift lever
  7. Sleeve
  8. Spring
  9. Snap ring
10. Pin
11. Tower cover
12. Fill plug
13. Forward/Reverse shift lever
14. Cover
15. Retaining nut
16. Pin
17. Sleeve
18. Top cover
19. High/Low shift lever

Fig. 210—Forward/Reverse shift lever (13—Fig. 209) is operated by linkage similar to type shown.

Fig. 211—Adjust lengths of shift control rods attached to the relay lever (R) to provide alignment of forward/reverse shift linkage as shown.

Fig. 212—Exploded view of shift rails and forks for eight-speed shuttle transmission.

1. Shift rail (High/Low)	10. Shift fork
2. Shift gate	11. Shift gate
3. Shift fork	12. Plate
4. Detent springs and plungers	13. Gasket
5. Retaining screws	14. Interlock plunger & balls
6. Shift rail (Forward/Reverse)	15. Retainer
7. Shift gate	16. Shift rail (3rd/4th)
8. Shift fork	17. Shift gate
9. Shift rail (1st/2nd)	18. Shift fork

sion top cover. The reason for removing top cover will determine what additional procedures are necessary.

"Loctite 515 Instant Gasket" or equivalent should be used to seal tower to top cover. Tighten retaining screws to 50-70 N·m (37-52 ft.-lbs.) torque.

SHIFTER RAILS AND FORKS

Models With Eight-Speed Shuttle Transmission

210. To remove the shifter rails and forks, first refer to paragraph 158 or paragraph 159 and separate tractor between rear of transmission and front of axle center housing. Refer to paragraph 209 and remove the shift lever and tower assembly and transmission top cover. Models without cab must have instrument and steering console removed before top cover can be unbolted and removed.

On all models, remove the neutral safety start switch (N—Fig. 213). Remove safety wire and set screws (5—Fig. 212) and lift detent springs and plungers (4) from bores in housing. Unbolt and remove interlock parts (12, 13, 14 and 15—Fig. 213). Slide rail out of housing toward rear and remove gate and shift fork through top. Twist rails if necessary

while withdrawing. Fork (8—Fig. 212) can not be removed until transmission input shaft and gears are removed as outlined in paragraph 214.

When assembling, lubricate rails and make sure that rails do not have burrs, especially around the set screw locations, that would prevent easy and smooth installation. Each of the four shift rails is different

**Fig. 213—Drawing of shift rails, installed. Notice location
of neutral start switch (N). Interlock balls (14) are located
in retainer (15). Refer to Fig. 212 for legend.**

PLANETARY UNIT

Models With Eight-Speed Shuttle Transmission

211. To remove the planetary unit, first refer to
paragraph 158 or paragraph 159 and separate tractor
between rear of transmission and front of axle center
housing. Remove set screw (5—Fig. 212) from fork (3),
then remove the fork and coupler from rear of planetary. Remove the four retaining cap screws and withdraw rear cover (13—Fig. 214), rear thrust ring (6)
and planet carrier (7). Pry planetary ring gear (3)
with dowels from rear of housing. Remove planetary
front cover (2) and shim (1).

To disassemble planet carrier (7), press planet pins
(8) forward out of carrier. Remove planet gears (10)
with needle rollers (12) and thrust washers (9).
Splined sleeve (15) is retained in hub of carrier (7) by
retaining ring (16).

When reassembling, be sure to account for all of the
needle rollers (12). Each pinion contains two paths of
rollers separated by a washer (11) and each roller
path contains 27 rollers (12). Use petroleum jelly (not
heavy grease) to hold rollers in place while pressing
planet pins into carrier.

Be sure that front plate (2) and rear plate (13) are
positioned with grooved side toward planet carrier.
Slots in front plate and shim (1) must be aligned and
toward top when installed on transmission. Use petroleum jelly to hold thrust rings (6) in place. Make

from the other rails. Tighten screws retaining the
interlock retainer (15—Fig. 212) to 40-47 N•m (30-35
ft.-lbs.) torque and the seven set screws (5) to 34-52
N•m (25-38 ft.-lbs.) torque. Install safety wire
through set screws (5) after tightening to prevent
loosening. Adjust position of shift fork (18) as follows.

Use special tools (MF 414/2 and MF 414/3 or
equivalent) to hold shift rails (6, 9 and 16—Fig. 212
and Fig. 213) in neutral detent position. Insert centralizing tool (MF 414/1 or equivalent) into holes in
shift forks (8, 10 and 18—Fig. 212) and into matching
hole of sliding synchronizing coupling. Tighten each
set screws (5S), while turning centralizing tool, to
hold the fork in this center position. Screws (5 and
5S) should be tightened to 34-52 N•m (25-38 ft.-lbs.)
torque and locked in position with safety wire. Shift
all gears, rails and gates to neutral after checking to
make sure that all shift gates align. Shift into each
gear to make sure that each can be correctly engaged.
Install detent plungers and springs (4—Fig. 212).
Remainder of assembly is the reverse of disassembly
procedure.

**Fig. 214—Exploded view of high/low range planetary unit
used on 340, 350, 355 and 360 models.**

1. Front shim	10. Pinion
2. Front plate	11. Spacer washer
3. Ring gear	12. Needle rollers (27/path)
4. Dowel	13. Rear plate
6. Thrust washer	14. Shift coupler
7. Planet carrier	15. Splined sleeve
8. Pinion shaft	16. Retaining ring
9. Thrust washer	17. Transmission mainshaft

certain that tangs of thrust rings engage notches in pinion carrier and that brass side faces away from carrier. Cutaway section of rear plate (13) should be positioned at the lower, **right** corner of planetary as shown. A lockwasher is used only on mounting screw at corner of cutaway. Tighten mounting screws to 40-47 N•m (30-35 ft.-lbs.) torque. Complete assembly procedure by reversing disassembly.

PTO INPUT SHAFT AND RETAINER HOUSING

Models With Eight-Speed Shuttle Transmission

212. To remove the pto input shaft (8—Fig. 215) and retainer (3), first split tractor between engine and transmission as outlined in paragraph 152. Remove clutch release bearing (16), retainer (15), fork (13) and release shaft (12). Remove brake cross shaft, if so equipped. Remove screws and pull pto input shaft (8) and retainer (3) as a unit from transmission housing.

To disassemble, remove snap ring (6) from rear of retainer, then push input shaft and bearing (5) out of housing toward rear. Bearing can be removed from input shaft after removing snap ring (4). Remove oil seals (1 and 9) and needle bearing (2) only if new parts are to be installed.

Lubricate all seals and bearings with petroleum jelly before assembling. Use special tool (MF 315A or equivalent) to install new needle bearing (2). Install seal (9) with lip toward rear, using special tool (MF 421 or equivalent). Install ball bearing (5) on shaft with shield toward rear, then install snap ring (4). Use special tool (MF255B or equivalent) to install seal (1) with lip toward rear. Use special sleeve (KMF

1004 or equivalent) to protect seal (1), slide input shaft into the retainer housing, then install snap ring (6). Install "O" ring (7) and install housing and shaft assembly over transmission input shaft (11) into housing bore. Coat threads of retaining screws with "Hylomar" or equivalent sealer and tighten to 54-61 N•m (40-45 ft.-lbs.) torque. Remainder of assembly is reverse of disassembly.

PTO LOWER SHAFT FRONT BEARING AND RETAINER

Models With Eight-Speed Shuttle Transmission

213. The pto lower shaft front bearing (41—Fig. 216) can be removed and installed after separating tractor between the rear of the engine and front of transmission housing. Removal of the lower driven gear (45) and shaft (46) require the additional steps of separating at rear of transmission housing and removal of the transmission top cover and shift rails. The following procedure describes only service to the front bearing and bearing retainer (42) after splitting as outlined in paragraph 152.

Remove clutch release bearing (16—Fig. 215), retainer (15), fork (13) and release shaft (12). Unbolt and remove cover (36—Fig. 216), then remove snap ring (38) and washer (39). Thread a 3/8 inch UNC cap screw, 75 mm (3 inches) long into each of the two threaded holes of retainer (42). Tighten the two jackscrews to push retainer (42) and bearing (41) forward off shaft (46). Bearing can be pressed from retainer after removing snap ring (40). Remove "O" ring (37) and gasket (43), then clean all parts, thoroughly.

Coat gasket (43) and threads of retaining screws lightly with "Hylomar" or equivalent sealer before

Fig. 215—Pto input shaft (8) and retainer housing (3) can be removed as a unit from front of transmission housing.

1. Seal
2. Needle bearing
3. Retainer housing
4. Snap ring
5. Ball bearing
6. Snap ring
7. "O" ring
8. Pto input shaft
9. Oil seal
10. Thrust washer
11. Transmission input shaft
12. Clutch lever and shaft
13. Clutch release lever
14. Return spring
15. Release housing
16. Release bearing

Fig. 216—Exploded view of the pto lower shaft and bearings.

36. Cover
37. "O" ring
38. Snap ring
39. Washer
40. Snap ring
41. Ball bearing
42. Retainer
43. Gasket
44. Snap ring
45. Pto lower gear
46. Pto lower shaft

installing. End of shaft (46) is threaded to accept a puller screw to pull bearing (41) onto shaft until washer (39) and snap ring (38) can be installed. Align holes in retainer (42) with holes in transmission housing, install "O" ring (37) and cover (36). Tighten retaining screws to 54-61 N·m (40-45 ft.-lbs.) torque. Remainder of assembly is reverse of disassembly procedure.

TRANSMISSION INPUT SHAFT

Models With Eight-Speed Shuttle Transmission

214. To remove the transmission input shaft (11—Fig. 217 and Fig. 218), first remove the transmission as outlined in paragraph 208. Remove the top cover and shift levers as outlined in paragraph 209, planetary unit as outlined in paragraph 211 and the shift rails and forks as outlined in paragraph 210. Refer to paragraph 212 and remove the pto input shaft and retainer. Slide input shaft (11—Fig. 218) forward, then withdraw gears (81 and 82) out top. Forward/reverse shift fork (8—Fig. 212) can be removed after removing input gears.

Reinstall input shaft by reversing removal procedures. Refer to paragraph 210 for adjusting neutral position of forks on shift shafts.

MAINSHAFT (OUTPUT SHAFT) AND GEARS

Models With Eight-Speed Shuttle Transmission

215. To remove the mainshaft (19—Fig. 218), first remove the transmission as outlined in paragraph

208. Remove the top cover and shift levers as outlined in paragraph 209, planetary unit as outlined in paragraph 211 and the shift rails and forks as outlined in paragraph 210. Refer to paragraph 212 and remove the pto input shaft and retainer, then remove the transmission input shaft as outlined in paragraph 214.

Remove spacer (83—Fig. 218), lift snap ring (14) from groove, push mainshaft (19) to rear and remove gear (13). Continue to move mainshaft toward rear while lifting synchronizer assembly (90 through 95), thrust washer (61), gears (20, 21 and 24) and synchronizer assembly (63 through 68) out top opening. Rear bearing (22) can be removed if necessary after removing snap ring (23).

Synchronizer parts (64 through 67 and 90 through 95) should be removed as an assembly and separated only if necessary. Wrap synchronizer assembly with a cloth to catch poppet parts (65 and 66 or 77 and 78) when sliding coupling (64 or 76) is removed from hub (67 or 79). Clean, inspect and renew any components that are damaged or worn excessively. Place synchronizer cones (72 and 73 or 89 and 96) and synchronizer rings (63 and 68 or 90 and 95) on respective gears (21 and 24 or 13 and 20). Make sure that each ring is seated squarely on taper of cone. Measure clearance between ring and gear at several places around gear, using a feeler gauge. Install new synchronizer ring (63, 68, 90 or 95) if clearance is less than 0.5 mm (0.020 inch). This clearance can be checked with gearbox assembled, but minimum clearance is then increased to 0.8 mm (0.030 inch).

When reassembling the synchronizer shift components, align centralizing holes (H—Fig. 219) in shift collar (64 or 91) and hub (67 or 92). Special tools (MF 415 and MS 550) are available to assist holding the three poppets (65 or 93) into hub (67 or 92), against spring pressure, while installing coupling (64 or 91). Synchronizer rings (75 and 80) on countershaft are

Fig. 217—Cross section of eight-speed shuttle transmission.

different than similar parts on mainshaft, because countershaft rotates in opposite direction.

Make sure that forward and reverse shift fork (8—Fig. 212) is in place in shift collar before installing the mainshaft.

Reassemble mainshaft (19—Fig. 218), snap rings (14 and 23), bearings (15 and 22) and all parts between snap rings (14 and 23), outside of gearbox housing. Be sure that sleeves (59, 60, 62 and 70) match respective gears (13, 20, 21 and 24) in length. Refer to Fig. 217 for cross section of assembled transmission. Measure clearance between front snap ring (14—Fig. 218) and inner race of bearing (15). Select spacer (17) of proper thickness to limit gap to 0.080-0.300 mm (0.003-0.012 inch). Spacer (17) is available in four thicknesses of 4.14-4.19, 4.39-4.44, 4.62-4.67 and 4.85-4.90 mm (0.163-0.165, 0.173-0.175, 0.182-0.184 and 0.191-0.193 inch). Reassemble mainshaft and components inside housing using the preselected spacer (17). Flat side of spacer (17) should be toward gear (13).

Remainder of assembly is reverse of disassembly procedure. Coat threads of screws attaching retainer housing (3) lightly with "Hylomar" or equivalent sealer before tightening to 54-61 N·m (40-45 ft.-lbs.) torque.

COUNTERSHAFT (LAYSHAFT) AND GEARS

Models With Eight-Speed Shuttle Transmission

216. The mainshaft must be removed as outlined in paragraph 215 and pto lower shaft (46—Fig. 218) and drive gear (45) should be removed as outlined in paragraph 213, before removing countershaft (31). A $7/16$ inch UNF screw approximately 75-100 mm (3-4 inches) long can be threaded into front of pto lower shaft and used to assist in removal of lower pto shaft. Remove snap ring (25) from rear, bump shaft forward enough to remove snap ring (29) from groove and move snap ring forward to unsplined part of countershaft. Remove snap ring (35) and thrust washer (84). Drive countershaft (31) to the rear and slide reverse gear (34) forward as far as possible. Remove snap ring (86) from groove, then slide snap ring forward as far as possible. Drive countershaft (31) toward rear as far as possible and remove reverse gear (34). Remove snap ring (86) and synchronizer ring (75), then remove synchronizer assembly (76 through 79) as a unit. Slide synchronizer ring (80), synchronizer cone (85), gear (33) and washer (88) from shaft. Drive shaft and bearing (32) forward through center web, then slide shaft (31), snap ring (87) and bearing (32) out toward front and remove gears (27, 28 and 69).

Synchronizer parts (76 through 79) should be removed as an assembly and separated only if necessary. Wrap synchronizer assembly with a cloth to catch poppet parts (77 and 78) when coupling (76) is removed from hub (79). Clean, inspect and renew any

components that are damaged or worn excessively. Place synchronizer cones (74 and 85) and synchronizer rings (75 and 80) on respective gears (33 and 34). Make sure that each ring is seated squarely, then measure clearance between ring and gear at several places around gear, using a feeler gauge. Install new synchronizer ring if clearance is less than 0.5 mm (0.020 inch). This clearance can be checked with

Fig.218—Exploded view of eight-speed shuttle transmission. Synchronizer (63 through 68) is made up of same components as synchronizer (75 through 80).

3. Retainer
8. Pto input shaft and gear
11. Transmission input shaft
12. Spacer
13. Gear (1st)
14. Snap ring
15. Bearing
16. Snap ring
17. Spacer (Variable thicknesses)
18. Roller bearing
19. Mainshaft
20. Gear (2nd)
21. Gear (4th)
22. Ball bearing
23. Snap ring
24. Gear (3rd)
25. Snap ring
26. Ball bearing
27. Gear (3rd)
28. Gear (4th)

29. Snap ring
30. Roller bearing
31. Countershaft & 1st gear
32. Ball bearing
33. Forward gear
34. Reverse gear
35. Snap ring
44. Snap ring
45. Pto driven gear
46. Pto lower shaft
50. Thrust washers
51. Washers
52. Spacer
53. Needle rollers
54. Gear (Reverse idler)
56. "O" ring
57. Idler shaft
59. Sleeve
60. Sleeve
61. Washer

62. Sleeve
63. Synchronizer ring (Same as 68)
64. Shifting coupler
65. Pressure blocks (3 used)
66. Ball pins & springs (3 used)
67. Synchronizer hub
68. Synchronizer ring (Same as 63)
69. Gear (2nd)
70. Sleeve
71. Washer
72. Synchronizer cone
73. Synchronizer cone
74. Synchronizer cone
75. Synchronizer ring (Same as 80)
76. Shifting coupler
77. Pressure blocks (3 used)

78. Ball pins & springs (3 used)
79. Synchronizer hub
80. Synchronizer ring (Same as 75)
81. Forward gear
82. Reverse gear
83. Spacer
84. Thrust washer
85. Synchronizer cone
86. Snap ring
87. Snap ring
88. Thrust washer
89. Synchronizer cone
90. Synchronizer ring
91. Shifting coupler
92. Hub
93. Pressure blocks (3 used)
94. Ball pins & springs (3 used)
95. Synchronizer ring
96. Synchronizer cone

Fig. 219—Synchronizer rings for unit used on countershaft are different than similar parts used on mainshaft. Holes (H) are used when checking and adjusting position of shift forks and must be aligned when assembling synchronizer. Refer to Fig. 218 for legend.

gearbox assembled, but minimum clearance is increased to 0.8 mm (0.030 inch).

When reassembling the synchronizer shift components, align centralizing holes (H—Fig. 219) in shift collar (76) and hub (79). Synchronizer rings (75 and 80) on countershaft are different than similar parts on mainshaft, because countershaft rotates in opposite direction. Special tools (MF 415 and MS 550) are available to assist holding the three poppets (77 and 78) into hub (79) while installing coupling (76).

Reassemble by reversing removal procedure. Always install new snap rings when reassembling. Position snap ring (29) around shaft in the unsplined area and install front bearing (32) on shaft before

installing shaft in housing. Refer to Fig. 217 for cross section of assembled transmission. Refer to paragraph 215 for assembly of mainshaft and selection of thrust washer (17—Fig. 218).

Coat gasket (43—Fig. 216) lightly with "Hylomar" or equivalent sealer before installing. Coat threads of screws attaching cover (36) and screws attaching retainer housing (3—Fig. 215) lightly with "Hylomar" or equivalent sealer before tightening to 54-61 N·m (40-45 ft.-lbs.) torque.

REVERSE IDLER GEAR AND SHAFT

Models With Eight-Speed Shuttle Transmission

217. To remove reverse idler gear (54—Fig. 218), shaft (57) and related parts, first remove the transmission as outlined in paragraph 208. Remove the top cover and shift levers as outlined in paragraph 209, planetary unit as outlined in paragraph 211 and the shift rails and forks as outlined in paragraph 210. Refer to paragraph 213 and remove the pto lower shaft front bearing and retainer, then refer to paragraph 214 and remove the transmission input shaft. Remove the spacer (83) from front of mainshaft.

Carefully withdraw reverse idler shaft (57) from front and lift idler gear (54), spacer (52), washers (51), thrust washers (50) and bearing rollers (53) through top opening, being careful not to drop any parts into housing. Each bearing path (53) contains 23 rollers.

Assemble needle rollers (53) and washers (51) in reverse idler gear using petroleum jelly. Position thrust washers (50), spacer (52) and gear and bearing assembly in housing, then install idler shaft (57), using new "O" ring (56). Complete reassembly by reversing disassembly procedure.

TRANSMISSION (TWELVE-SPEED SHUTTLE)

399 Models

Twelve forward speeds are possible by using a synchromesh gearbox with four forward speeds compounded by a three-speed range unit. An equal number of reverse speeds is made possible by fitting a forward/reverse unit in the front of gearbox. Refer to Fig. 220 for shift pattern of twelve-speed shuttle transmission available in 399 model tractors. The forward/reverse shift lever (1) is located to the left of the steering wheel and can be moved when the clutch

pedal is depressed. The engine can be started only when forward/reverse lever is in neutral (center) position. Selection of the twelve speed ranges is accomplished by moving a single lever (2). The shift lever is spring loaded to the center, 3rd and 4th speed, position. A three-speed range transmission is shifted by moving lever (2) to neutral position, then to the far right, before moving the ratcheting shift either up or down. An indicator (3) shows which range is selected. After range is selected, the shift lever can be moved to select one of the four gears of the main transmission in a conventional manner.

Fig. 221—View of left side showing hose (2) and solenoid (1) to operate the four-wheel drive. The range indicator switch (3) is retained by three screws.

Fig. 220—Shift pattern for twelve-speed shuttle transmission. The forward/reverse shift lever is located at (1) on left of steering wheel. Selection of the twelve speed ranges is accomplished by moving a single lever (2). Indicator (3) shows which of the three ranges is engaged.

LUBRICATION

Models With Twelve-Speed Shuttle Transmission

218. The transmission, rear differential and hydraulic system share common fluid which is contained in the transmission and center housing. Recommended lubricant is Massey-Ferguson Super 500 multi-use 10W-30 oil, Massey-Ferguson Permatran or equivalent transmission/hydraulic oil. Capacity is 47.4 L (12.5 gal.). Tractors are equipped with dipstick with marks indicating maximum and minimum oil levels. Level should be maintained at maximum level if tractor is operated on hilly terrain or if using implements which require large quantities of oil.

TRACTOR REAR SPLIT

Models With Twelve-Speed Shuttle Transmission

219. Do not attempt to split tractor between the range unit and the main gearbox. The main gearbox

and range gearbox should be separated only after they have been removed from the tractor as a unit.

To separate tractor between the rear of the transmission housing and the front of the rear axle center housing, set parking brake and block the wheels to prevent rolling. Disconnect battery ground cable and drain all transmission fluid.

If equipped with four-wheel drive, refer to paragraph 9 and remove the front drive shaft. Unscrew and remove the solenoid (1—Fig. 221) and detach hydraulic line (2). Detach four-wheel-drive indicator switch and four-wheel-drive selector lever.

On models with cab, disconnect throttle stop cable and throttle cable from fuel injection pump and move cables out of the way so that cables will not catch when separating tractor. Mark (identify), disconnect and cap steering hoses, heater hoses and air conditioning lines (if so equipped), which would interfere. Remove front floor mat and floor plates.

On models without cabs, remove floor mats, floor panels, lower (gearbox) cover, both foot steps and battery boxes.

On all models, unbolt and withdraw range indicator unit (3), disconnect clutch linkage and remove foot throttle pedal. Remove knob (1—Fig. 222) from main shift lever, then remove the metal guard (2) from around shift linkage. Remove the lower strut (3) from the shift lever, remove the shift lever pivot roll pin (4), then withdraw shift lever (5). Disconnect hydraulic oil supply pipe located on right side of tractor and, if so equipped, detach hydraulic pipes from left side. Disconnect forward/reverse shift linkage and brake lines, then drain brake system. Disconnect wiring harness at the connectors below console on right side and disconnect wires from safety start switch. Remove the hydraulic suction filter assembly. Place hardwood wedges between the front axle and axle

Fig. 222—Shift lever and related parts used with twelve-speed shuttle transmission.

1. Shift knob
2. Shield
3. Lower strut

4. Pivot roll pin
5. Shift lever

support casting on both sides to prevent tipping around axle pivot when tractor is separated. Support rear of tractor under axle center housing and front of tractor under transmission housing. Remove screws attaching rear axle center housing to four-wheel-drive transfer case, then separate the tractor halves. Refer to Fig. 223. Either the front can be moved forward or the rear can be moved to the rear, but it is important to support both halves safely and securely.

Install at least two dowel studs to facilitate alignment, position new gasket on dowel studs, then move tractor together, turning engine to align splines of shear tube. Make sure that flanges are tight against each other before installing retaining screws. Tighten all screws retaining the rear axle center housing to the transmission housing to 102-122 N•m (75-90 ft.-lbs.) torque, beginning at top center and progressing in a clockwise direction (as viewed from tractor rear) two times around the mating flange. If equipped with four-wheel drive, coat threads of drive shaft retaining screws with "Loctite 270," then attach drive shaft flanges and tighten screws to 55-75 N•m (40-55 ft.-lbs.) torque. Remainder of assembly is reverse of disassembly.

TRANSMISSION REMOVAL

Models With Twelve-Speed Shuttle Transmission

220. Refer to paragraph 219 and separate transmission from rear axle center housing, then support the gearbox assembly independently from the engine and front assembly. If not equipped with cab, unbolt and support or remove the steering and instrument console. On all models, remove screws and stud nuts

Fig. 223—View of tractor with standard cab separated between the transmission housing and the rear axle center housing. Special splitting stand and track is shown. Support and splitting procedures are similar for other models.

attaching transmission housing to the rear of engine, then separate transmission from engine.

Reassemble transmission to engine by reversing the splitting procedure and observe the following. Install a M12 alignment stud approximately 100 mm (4 inches) long in each side of the transmission housing to assist in aligning engine to transmission. Turn flywheel to align clutch plate splines with transmission and pto input shaft splines. Install retaining screws and nuts after engine and transmission housing flanges are completely together. Tighten retaining screws and nuts to 100-130 N·m (74-94 ft.-lbs.) torque.

Refer to paragraph 219 and reconnect transmission to rear axle center housing. Tighten all screws retaining the transmission housing to spacer housing, transfer case or rear axle center housing to 102-122 N·m (75-90 ft.-lbs.) torque, beginning at top center and progressing in a clockwise direction (as viewed from tractor rear) two times around the mating flange. On four-wheel-drive models, coat threads of drive shaft retaining screws with "Loctite 270," then attach drive shaft flanges and tighten screws to 55-75 N·m (40-55 ft.-lbs.) torque. On all models, check and adjust clutch linkage as outlined in paragraph 151.

Refer to paragraph 20 for removal of the four-wheel-drive transfer case from the range housing.

TOP COVER

Models With Twelve-Speed Shuttle Transmission

221. If transmission is not already removed or the top is not yet uncovered, remove the cab floor or foot plates. Disconnect the forward/reverse shift linkage. If equipped with cab, remove the three screws from the oil filler tube and remove the tube (6—Fig. 224). On all models, unbolt and remove top cover from transmission housing.

Install new "O" ring (4) if forward/reverse shift lever (1) is removed. Use thread lock such as "Loctite 270" on threads of set screw (2).

Before installing top cover, coat sealing surface with "Loctite 515" or equivalent sealer. Tighten retaining screws to 101-122 N·m (75-90 ft.-lbs.) torque.

MAIN GEAR SHIFT LEVER AND SELECTOR SHAFT

Models With Twelve-Speed Shuttle Transmission

222. Remove knob (1—Fig. 222) from main shift lever, then remove the metal guard (2) from around shift linkage. Remove the lower strut (3) from the

Fig. 224—View of transmission top cover and forward/reverse shift lever.

N. Neutral switch
1. Forward/reverse shift
2. Set screw
3. Internal shift lever
4. "O" ring
5. Bushing
6. Fill tube

shift lever, remove the shift lever pivot roll pin (4), then withdraw shift lever (5).

Remove the two 12-point screws (11—Fig. 225) and withdraw the selector assembly (1 through 12) from the range housing.

Install new "O" ring (4) if unit is disassembled. Coat sealing surface of range housing and sealing surface of end cover (5) with "Loctite 515" or equivalent before installing. Tighten the 12-point screws (11) to 50-70 N·m (38-52 ft.-lbs.) torque.

FORWARD/REVERSE SHIFT CONTROL LINKAGE

Models With Twelve-Speed Shuttle Transmission

223. Refer to Fig. 226 and Fig. 227 for forward/reverse shuttle shift linkage. Be sure that electrical wiring does not touch any part of shift linkage. Lubricate linkage with general purpose grease, when assembling. Adjust lengths of rods attached to the relay lever as shown in Fig. 228. Install shims between relay lever pivot and retaining snap ring to remove any end play that would result in lost motion.

Fig. 225—Exploded view of the gear selector shaft and related parts.

1. Boot
2. Snap ring
3. "O" ring
4. Screws (6 point)
5. End cover
6. Spacer
7. Snap ring
8. Washers (2 used)
9. Spring
10. Support
11. Screws (12 point)
12. Selector shaft
13. Rear selector mechanism

Fig. 226—View of forward/reverse shift lever and linkage installed, but with steering wheel and lower cover removed.

MAIN TRANSMISSION SHIFT RAILS AND FORKS

Models With Twelve-Speed Shuttle Transmission

224. To remove the shift rails and forks for the main transmission, first remove the gearbox from the tractor as outlined in paragraph 220 and remove the top cover as outlined in paragraph 221. Refer to paragraph 222 and remove the main gear shift lever and

Fig. 227—Exploded view of forward/reverse shift linkage.

1. Shift lever
2. Link
3. Relay shaft & lever
4. Lever
5. Link
6. Clutch interlock linkage

selector shaft. Set the transmission on front end, with engine/clutch flange down. Unbolt range transmission housing from the main transmission housing, then carefully lift range gearbox from main gearbox.

Fig. 228—Adjust lengths of shift control rods attached to the relay lever (R) to provide alignment of forward/reverse shift linkage as shown.

Fig. 229—The rear selector mechanism can be removed after removing the cap screw from the left and the nut from the right.

Fig. 230—Exploded view of shift forks. Detent spring and plunger (4S), in the center, is different than detents and springs (4) on either side.

1. Range selector shaft	10. Shift fork
2. Cap screw	11. Return spring
3. Screw and plate	12. Support
3N. Nut	13. Gasket
4. Detent springs & plungers	14. Interlock plunger & balls
5. Retaining screws	15. Retainer
6. Shift rail (3rd/4th)	16. Shift rail (1st/2nd)
7. Roll pin	17. Shift gate
8. Shift fork	18. Shift fork
9. Shift rail (Forward/ Reverse)	

Unbolt and remove the rear selector mechanism (Fig. 229). Remove safety wire and set screws (5—Fig. 230) and lift detent springs and plungers (4) from bores in housing. Slide rail out of housing toward rear and remove gate and shift fork through top. Twist rails if necessary while withdrawing. Fork (8) can not be removed until transmission input shaft and gears are removed as outlined in paragraph 231.

When assembling, lubricate rails and make sure that rails do not have burrs, especially around the set screw locations, that would prevent easy and smooth installation. Each of the three shift rails is different from the other rails. Adjust position of shift forks (8, 10 and 18) as follows.

Use special tool MF 414/2 or equivalent (T2—Fig. 231) to hold shift rail (9—Fig. 230) in neutral detent

Fig. 231—View of special tool (T2) holding the rail in center detent position, while centering forward/reverse shift fork with special tool (T1).

position. Insert centralizing tool, MF 414/1 or equivalent, (T1—Fig. 231) into hole (H—Fig. 230) in forward and reverse shift fork (8), into matching hole of sliding synchronizing coupling and into hole in synchro-

Fig. 232—View of special tool (T2) holding the rail in center detent position, while centering shift fork with dial indicator (D).

Fig. 233—View of detents for shift rails. Note that center detent and spring (4S) are different than the other two (4).

nizer hub. Tighten each of the two set screws (5), while turning centralizing tool. Screws (5) should be tightened to 34-52 N·m (25-38 ft.-lbs.) torque and locked in position with safety wire.

To adjust the position of the 1st/2nd shift fork (18—Fig. 230) and 3rd/4th shift fork (10), proceed as follows. Use special tool MF 414/2 or equivalent (T2—Fig. 232) to hold shift rails (6 and 16—Fig. 230) in neutral detent position. Make sure that synchronizer is in neutral and is not engaging either gear, then mount a dial indicator with a magnetic base as shown at (D—Fig. 232) against the shift fork. Gently move the shift fork to the rear, then forward without engaging either gear, while noting the location of the fork as indicated by the dial indicator (D). Position the fork in exact center and tighten each of the two set screws (5). Screws should be tightened to 34-52 N·m (25-38 ft.-lbs.) torque and locked in position with safety wire. Make sure that tightening the screws does not move the position of the fork on rail. Use similar procedure to adjust other fork (10 or 18—Fig. 230) on rail (6 or 16).

Shift into each gear to make sure that each can be correctly engaged. Shift all gears, rails and gates to neutral after checking to make sure that all shift gates align. Install detent plungers and springs (4 and 4S—Fig. 233). Remainder of assembly is the reverse of disassembly procedure.

Reattach range gearbox to main transmission, using a new gasket and coating threads of retaining screws with "Hylomar" or equivalent sealer. After lowering range gearbox onto main gearbox, make sure that range shifter fork is correctly engaged. Tighten screws attaching range gearbox to main gearbox to 105 N·m (75 ft.-lbs.) torque. Refer to para-

graph 226 for selecting shims (40—Fig. 234) as required for setting shaft end play.

RANGE CHANGE UNIT UPPER SHAFT

Models With Twelve-Speed Shuttle Transmission

225. To remove the upper shaft from the range transmission, first remove the gearbox from the tractor as outlined in paragraph 220. Remove the main gear shift lever and selector shaft as outlined in paragraph 222, unbolt range transmission housing from the main transmission housing, then separate the two housings. Support the transmission on a suitable stand with rear of range gearbox down and enough clearance to withdraw shaft (37—Fig. 234). Remove screws (42) from front of selector rail, remove screw (43) from rear, then withdraw rail (45) from shifter forks (46 and 47). Support gears and forks using special tool MF477 (P2—Fig. 235) or equivalent. Attach a fixture such as MF476 (P1) and push the upper drive shaft out of bearing (14—Fig. 234).

CAUTION: Do not drop parts as shaft is withdrawn from bearing. Also, do not use a hammer and punch to remove the shaft from bearing or to install bearing on shaft, because serious, but undetected damage will occur.

Fig. 234—Exploded view of range section of twelve-speed shuttle transmission. Three sets of pins, springs and retainers (25) are located in each gear (24 and 34).

1. Boot	19. Springs and plungers	34. Synchronizer cone	51. Lower shaft
2. Snap rings	20. Synchronizer hub	35. Low range gear	52. Bearing cap
3. "O" ring	21. Shift collar (High range)	36. Bushing	53. Shims
4. Screws (6 point)	22. Drive gear (four-wheel drive)	37. Upper drive shaft	54. Tapered roller bearing
5. End cover	23. Bushing	38. Thrust washer	55. Thrust washer (2 pieces)
6. Spacer	24. Middle range gear	39. Tapered roller bearing	56. Middle range gear
7. Snap ring	25. Alignment pins, springs & retainers	40. Shims	57. Spacer
8. Washers (2 used)	26. Pressure blocks	41. Bearing cap	58. Driven gear
9. Spring	27. Springs and plungers	42. Screws	59. Tapered roller bearing
10. Support	28. Synchronizer cone	43. Screw	60. Indicator switch cover
11. Screws (12 point)	29. Synchronizer ring	44. Plates (2 used)	61. Range indicator switch
12. Selector shaft	30. Synchronizer hub	45. Selector rail	62. Roller bearings
14. Tapered roller bearing	31. Shift collar (Medium & Low range)	46. Shift fork (High range)	63. Spacer rings
15. Thrust washer	32. Synchronizer ring	47. Shift fork (Middle/Low range)	64. Drive gear (four-wheel drive)
16. High range synchronizer cone	33. Synchronizer bushing	48. Rollers and pins	65. Transfer case (four-wheel drive)
17. Synchronizer ring		49. Shifter cam assembly	66. Cover (two-wheel drive)
18. Pressure blocks		50. Bearing	

Remainder of disassembly procedure will be evident. Pins, springs and retainers (25—Fig. 234) can be driven from gears (24 and 35) as shown in Fig. 236 if renewal is required. Refer to paragraph 228 for removal and service to shifter cam assembly (49—Fig. 234).

Assemble synchronizer cone (28 or 34) and synchronizer ring (29 or 32) on appropriate gear (24 or 35) and measure gap with feeler gauge as shown in Fig. 237. If average clearance is less than 0.5 mm (0.020 inch), install new brass synchronizer ring.

Use petroleum jelly to hold synchronizer cones, rings, bearings and other parts together while assem-

bling. Install low range gear (35—Fig. 234), bushing (36) and related parts in place, using special holding fixture MF477 or equivalent (P2—Fig. 235). "V" notches in synchronizer cone (34) must engage drive pins (25) in gear (35). Assemble Middle/Low range synchronizer (29 through 32—Fig. 234) and shift fork (47) over Low drive gear (35). Pins and rollers (48) must be properly located on shift fork and must correctly engage track in cam (49). Be sure that locating lugs of synchronizer ring (29) engage slots of hub (30). Install Middle range gear (24) and bushing (23), making sure that "V" notches in synchronizer cone (28) engage drive pins (25) of gear (24). Install

Fig. 235—View of special tool MF476 (P1) attached for pushing the upper drive shaft out of front tapered bearing. Special tool MF477 (P2) is available to hold gears and synchronizers in position while removing and installing shaft.

Fig. 237—Measure gap (G) with a feeler gauge (F) as shown to check condition of synchronizer rings.

without shims (40) and insert shaft up through the installed gears, bushings, hubs and related parts. Install screws to hold cap (41) in place, but selection of shims (40) will require later removal of cap. Install thrust washer (15) with smaller diameter up (toward front), then press tapered roller bearing (14) onto end of shaft.

> **CAUTION: Do not use a hammer and punch to drive bearing onto shaft. Damage resulting in seizure and extensive failure to many components may occur.**

A special pressing fixture MF476 (P1—Fig. 235) is available to assist in removal and installation of bearing (14—Fig. 234). Make sure that bearing is fully in place by attempting to turn the thrust washer (15). Install shifter rail (45) and screws (42 and 43). Coat threads of screws (42 and 43) with "Loctite 270" or equivalent and tighten to 25-35 N·m (18-26 ft.-lbs.) torque. Rotate shafts (37 and 51) to make sure that all gears are free to rotate and are not binding.

Reattach range gearbox to main transmission, using a new gasket and coating threads of retaining screws with "Hylomar" or equivalent sealer. After lowering range gearbox onto main gearbox, make sure that range shifter fork is correctly engaged. Tighten screws attaching range gearbox to main gearbox to 105 N·m (75 ft.-lbs.) torque. Refer to paragraph 226 for selecting shims (40) as required for setting shaft end play.

Fig. 236—Synchronizer aligning pins, springs and retainers can be removed by driving from gears with a punch as shown.

four-wheel-drive gear (22) with grooved thrust face down toward gear (24). Assemble synchronizer (16 through 21) and High range shift fork (46) into position. Make sure that pins and rollers (48) correctly engage track in cam (49) and bushing bores in shift forks are aligned with bore for shift rail (45). Assemble thrust washer (38), bearing (39) and cap (41)

RANGE CHANGE UPPER SHAFT BEARING PRELOAD

Models With Twelve-Speed Shuttle Transmission

226. The selection of shims (40—Fig. 234) determines the adjustment of bearings (14 and 39) and proper adjustment of these bearings is critical to longevity of transmission parts. Some special tools are required. The special rotating tool (T3—Fig. 238) can be locally fabricated from an old shear tube or pipe and handle which is used to turn the transmission shaft before measuring gap with feeler gauges (F). Special tool MF478 (T4) is four special length M6 screws and four special springs which are used to load cap (41) to about 45 kg. (100 lbs.).

To check bearing preload, set transmission on floor with engine/clutch surface down and remove the four screws retaining bearing cap (41—Fig. 238) to the range gearbox housing. Install four M6 screws with springs (T4) and tighten the screws until the combined pressure of the four springs is approximately 45 kg. (100 lbs.) and gap between cap and housing is even. If special screws and springs MF478 are used, height (H) between top of screw heads and bearing cap is 55 mm (2.17 inches). Engage medium range by moving the range rear shift coupling down (toward front), then turn the upper shaft with tool (T3) approximately ten revolutions to make sure that bearings are correctly seated. Measure clearance between bearing cap and gearbox housing carefully in at least four locations around cap. If measurements are not the same around cap, rotate upper shaft again with special tool and make sure that special tool springs

Fig. 238—Refer to text for selecting shims (40—Fig. 234). Special tool (T4) is four M6 screws and four springs.

are all compressed the same amount, then remeasure. Subtract approximately 0.05 mm (0.020 inch) from the gap measured with feeler gauges, then install shims (40—Fig. 234) equal to this thickness. An example is a measurement of 0.37 mm - (minus) 0.05 mm equals 0.32 mm. Shims are available in thicknesses of about 0.5, 0.20 and 0.50 mm (0.002, 0.008 and 0.020 inch). Be sure to measure thickness of shims selected. Slightly less preload is required if original bearings are reinstalled.

Remove the special screws, springs, turning handle and bearing cap, then install shims determined to be correct thickness. Install bearing cap, coat threads of retaining screws with "Loctite 270" or equivalent and tighten retaining screws to 25-35 N·m (18-26 ft.-lbs.) torque. Disengage middle range gear and check both shafts for free rotation.

RANGE CHANGE LOWER SHAFT

Models With Twelve-Speed Shuttle Transmission

227. Refer to paragraph 225 and remove the upper shaft. Unbolt and remove the lower bearing cap (52—Fig. 234), shims (53) and cup for bearing (54). Withdraw lower shaft (51), cone for bearing (54) and thrust washer (55), then lift middle range gear (56), spacer (57), lower shaft drive gear (58) and if so equipped, the four-wheel-drive gear (64). Press cone of bearing (54) toward rear (gear end of shaft) to remove the two piece thrust washer (55).

Install cup for bearing (59) in housing bore, then install drive gear (58) with cone for bearing (59). On four-wheel-drive models, assemble spacer rings (63), bearings (62) and gear (64) over spacer (57), then position the unit in housing against gear (58). On two-wheel-drive models, install spacer (57) against gear (58). On all models, press bearing (50) into lower shaft and press cone for bearing (54) onto shaft. Position both halves of thrust washer (55) in groove of shaft, then press cone of bearing (54) toward front of shaft over thrust washer. Install middle range gear (56), then insert the lower shaft (51) through gears and spacers. Install bearing cap (52) and cone for bearing (54), using 1.0 mm (0.039 inch) thickness of shims (53). Shims are available in 0.05, 0.20 and 0.50 mm (0.002, 0.008 and 0.020 inch) thicknesses. Tighten screws retaining cap (52) to 25-35 N·m (18-26 ft.-lbs.) torque and mount a dial indicator as shown in Fig. 239 to measure shaft end play. Bearing preload should be 0.10 mm (0.004 inch) for new bearings or 0.05 mm (0.002 inch) for used bearings. Unbolt and remove cap (52), then reduce thickness of shims the amount of the measured end play plus the additional amount of desired preload. Reinstall cap with the required amount of shims, coat threads of retaining

Fig. 239—Mount a dial indicator (D) as shown, then pry gear up (P) to check end play of lower shaft in bearings.

screws with "Loctite 270" or equivalent and tighten screws to 25-35 N·m (18-26 ft.-lbs.) torque. Check shaft for free rotation, refer to paragraph 225 to install the upper shaft and to paragraph 227 for setting upper shaft bearings.

RANGE SELECTOR CAM

Models With Twelve-Speed Shuttle Transmission

228. Refer to paragraph 225 and remove the upper shaft and selector forks. Remove cover (60—Fig. 234)

and indicator switch (61), if not already removed. Remove plug (67), spring and plunger (68). Remove screw (69) and withdraw cam assembly (49) from housing. Refer to Fig. 240 for disassembled view of switch and cam assembly. Further disassembly is possible after removing snap ring.

Refer to Fig. 241 for correct assembly of pawls (79) and spring plate (77). Stick rollers (76) to cam block (75) with petroleum jelly, while assembling. Use a small screwdriver or similar tool to carefully guide pawls (79) into position so that pins (P) will enter holes in pawls. Coat threads of screw (69—Fig. 240) with "Loctite 270" or equivalent and tighten to 25-35 N·m (18-26 ft.-lbs.) torque. Install plunger and spring (68) and plug (69) making sure that selector cam is in neutral position. Tighten plug (69) to 50-70 N·m (38-52 ft.-lbs.) torque. Refer to paragraph 225 for installation of the shift forks and upper shaft.

PTO INPUT SHAFT AND RETAINER HOUSING

Models With Twelve-Speed Shuttle Transmission

229. To remove the pto input shaft (8—Fig. 242) and retainer (3), first split tractor between engine and transmission as outlined in paragraph 152. Remove clutch release bearing (16), retainer (15), fork (L) and release shaft (S). Remove brake cross shaft, if so equipped. Remove screws and pull pto input shaft (8) and retainer (3) from transmission housing as a unit.

Fig. 240—Exploded view of the range selector cam and the indicator switch shown in Fig. 234.

49. Shifter cam
60. Indicator switch cover
61. Indicator switch housing
67. Plug
68. Spring and detent
69. Screw
70. Indicator switch
71. Snap ring
72. Thrust washer
73. Needle bearings
74. Snap ring
75. Cam block
76. Rollers
77. Spring plate
78. "O" ring
79. Pawls
80. Spring

Fig. 241—View of selector cam, spring plate, ratchet pawls and spring correctly assembled.

To disassemble, remove snap ring (6) from rear of retainer, then push input shaft and bearing (5) out of housing toward rear. Bearing can be removed from input shaft after removing snap ring (4). Remove oil seals (1 and 9) and needle bearing (2) only if new parts are to be installed.

Lubricate all seals and bearings with petroleum jelly before assembling. Use special tool (MF 315A or equivalent) to install new needle bearing (2). Install seal (9) with lip toward rear, using special tool (MF 421 or equivalent). Install ball bearing (5) on shaft with shield toward rear, then install snap ring (4). Use special tool (MF255B or equivalent) to install seal (1) with lip toward rear. Use special sleeve (KMF 1004 or equivalent) to protect seal (1), slide input shaft into the retainer housing, then install snap ring (6). Install "O" ring (7) and install housing and shaft assembly over transmission input shaft (11) into housing bore. Coat threads of retaining screws with "Hylomar" or equivalent sealer and tighten to 54-61 N•m (40-45 ft.-lbs.) torque. Remainder of assembly is reverse of disassembly.

Fig. 242—Exploded view showing the pto and transmission input shafts and related parts. Clutch release lever is shown at (L) and lever shaft is shown at (S). A spacer is used in place of gear (82) on 399 models and smaller tractors with high speed shuttle transmission.

L. Clutch release lever	10. Needle bearing	15. Release housing
S. Clutch lever and shaft	11. Transmission input	16. Release bearing
1. Oil seal	shaft	18. Needle bearing
2. Needle bearing	12. Thrust washer	81. Forward gear
3. Retainer housing	13. Snap ring	82. Reverse gear
4. Snap ring	14. Return spring	83. Spacer
5. Ball bearing		
6. Snap ring		
7. "O" ring		
8. Pto input shaft		
9. Oil seal		

PTO LOWER SHAFT FRONT BEARING AND RETAINER

Models With Twelve-Speed Shuttle Transmission

230. The pto lower shaft front bearing (41—Fig. 243) can be removed and installed after separating tractor between the rear of the engine and front of transmission housing. Removal of the lower driven gear (45) and shaft (46) require the additional steps of separating at rear of transmission housing and removal of the transmission top cover and shift rails. The following procedure describes only service to the front bearing and bearing retainer (42) after splitting as outlined in paragraph 152.

Remove clutch release bearing (16—Fig. 242), retainer (15), fork (13) and release shaft (12). Unbolt and remove cover (36—Fig. 243), then remove snap ring (38) and washer (39). Thread a ⅜ inch UNC cap screw, 75 mm (3 inches) long into each of the two threaded holes of retainer (42). Tighten the two jackscrews to push retainer (42) and bearing (41) forward off shaft (46). Bearing can be pressed from retainer after removing snap ring (40). Remove "O" ring (37) and gasket (43), then clean all parts thoroughly.

Coat gasket (43) and threads of retaining screws lightly with "Hylomar" or equivalent sealer before installing. End of shaft (46) is threaded to accept a puller screw to pull bearing (41) onto shaft until washer (39) and snap ring (38) can be installed. Align holes in retainer (42) with holes in transmission housing, install "O" ring (37) and cover (36). Tighten retaining screws to 54-61 N.m (40-45 ft.-lbs.) torque. Remainder of assembly is reverse of disassembly procedure.

MAIN TRANSMISSION INPUT SHAFT

Models With Twelve-Speed Shuttle Transmission

231. To remove the transmission input shaft (11—Fig. 242), first remove the transmission as outlined in paragraph 220. Remove the top cover and shift levers as outlined in paragraph 221 and the shift rails and forks as outlined in paragraph 224. Refer to paragraph 229 and remove the pto input shaft and retainer. Slide input shaft (11—Fig. 242) forward, then withdraw gears (81 and 82) and spacer (83) out top. Be careful that bearing (18) does not fall into transmission when withdrawing shaft (11). A spacer is installed in place of gear (82) on 399 models. Forward/reverse shift fork (8—Fig. 230) can be removed after removing input gears.

Reinstall input shaft by reversing removal procedures. Refer to paragraph 224 for correctly setting position of forks on shift shafts.

MAIN TRANSMISSION UPPER SHAFT AND GEARS

Models With Twelve-Speed Shuttle Transmission

232. To remove the mainshaft (19—Fig. 244), first remove the transmission as outlined in paragraph 220. Remove the top cover and shift levers as outlined in paragraph 221 and the shift rails and forks as outlined in paragraph 224. Refer to paragraph 229 and remove the pto input shaft and retainer, then remove the transmission input shaft as outlined in paragraph 231.

Fig. 243—Exploded view of the pto lower shaft and bearings.

36. Cover
37. "O" ring
38. Snap ring
39. Washer
40. Snap ring
41. Ball bearing
42. Retainer
43. Gasket
44. Snap ring
45. Pto lower gear
46. Pto lower shaft

Lift snap ring (47—Fig. 244) from groove, push mainshaft (19) to rear and remove gear (13). Continue to move mainshaft toward rear while lifting synchronizer assembly (90 through 95), thrust washer (61), gears (20, 21 and 24) and synchronizer

assembly (63 through 68) out top opening. Rear bearing (22) can be removed if necessary.

Synchronizer parts (64 through 67 and 90 through 95) should be removed as an assembly and separated only if necessary. Wrap synchronizer assembly with

Fig. 244—Exploded view of twelve-speed shuttle transmission. Synchronizer (63 through 68) is made up of same components as synchronizer (75 through 80).

13. Gear (2nd)	35. Snap ring	63. Synchronizer ring	79. Synchronizer hub
19. Mainshaft	47. Snap ring	Same as 68)	80. Synchronizer ring
20. Gear (1st)	48. Bearing	64. Shifting coupler	Same as 75)
21. Gear (4th)	49. Spacer (Variable	65. Pressure blocks (3 used)	84. Thrust washer
22. Tapered roller bearing	thicknesses)	66. Ball pins & springs (3 used)	85. Synchronizer cone
23. Spacer	50. Thrust washers	67. Synchronizer hub	86. Snap ring
24. Gear (3rd)	51. Washers	68. Synchronizer ring	87. Snap ring
25. Snap ring	52. Spacer	(Same as 63)	88. Thrust washer
26. Ball bearing	53. Needle rollers (23/path)	69. Bushing	89. Synchronizer cone
& snap ring	54. Gear (Reverse	70. Sleeve	90. Synchronizer ring
27. Gear (3rd)	idler)	71. Washer	91. Shifting coupler
28. Gear (4th)	55. Bushing	72. Synchronizer cone	92. Hub
29. Spacer (2 piece)	56. "O" ring	73. Synchronizer cone	93. Pressure blocks (3 used)
30. Spacer	57. Idler shaft	74. Synchronizer cone	94. Ball pins & springs
31. Countershaft with	58. Bushings	75. Synchronizer ring	(3 used)
1st & 2nd gears	59. Sleeve	(Same as 80)	95. Synchronizer ring
32. Roller bearing	60. Sleeve	76. Shifting coupler	96. Synchronizer cone
33. Forward gear	61. Washer	77. Pressure blocks (3 used)	97. Bushing
34. Reverse gear	62. Sleeve	78. Ball pins & springs (3 used)	98. Bushing

a cloth to catch poppet parts (65 and 66 or 77 and 78) when sliding coupling (64 or 76) is removed from hub (67 or 79). Clean, inspect and renew any components that are damaged or worn excessively. Place synchronizer cones (72 and 73 or 89 and 96) and synchronizer rings (63 and 68 or 90 and 95) on respective gears (21 and 24 or 13 and 20). Make sure that each ring is seated squarely on taper of cone. Measure clearance between ring and gear at several places around gear, using a feeler gauge. Install new synchronizer ring (63, 68, 90 or 95) if clearance is less than 0.5 mm (0.020 inch). This clearance can be checked with gearbox assembled, but minimum clearance is then increased to 0.8 mm (0.030 inch).

When reassembling the synchronizer shift components, align centralizing holes (H—Fig. 245) in shift collar (64 or 91) and hub (67 or 92). Special tools (MF 415 and MS 550) are available to assist holding the three poppets (65 or 93) into hub (67 or 92), against spring pressure, while installing coupling (64 or 91). Synchronizer rings (75 and 80) on countershaft are different than similar parts on mainshaft, because countershaft rotates in opposite direction.

Make sure that forward and reverse shift fork (8—Fig. 230) is in place in shift collar before installing the mainshaft.

Reassemble mainshaft (19—Fig. 244), snap ring (47), bearings (48 and 22) and all parts between bearings (22 and 48), outside of gearbox housing. Be sure that sleeves (59, 60, 62 and 70) match respective gears (13, 20, 21 and 24) in length. Refer to Fig. 246 for cross section of assembled transmission. Measure clearance between front snap ring (47—Fig. 244) and inner race of bearing (48). Select spacer (49) of proper thickness to limit gap to 0.080-0.300 mm (0.003-0.012

inch). Spacer (49) is available in four thicknesses of 3.38-3.43, 3.63-3.68, 3.89-3.94 and 4.14-4.19 mm (0.133-0.135, 0.143-0.145, 0.153-0.155 and 0.163-0.165 inch). Reassemble mainshaft and components inside housing using the preselected spacer (49). Flat side of spacer (49) should be toward gear (13).

Remainder of assembly is reverse of disassembly procedure. Coat threads of screws attaching retainer housing (3—Fig. 242) lightly with "Hylomar" or equivalent sealer before tightening to 54-61 N·m (40-45 ft.-lbs.) torque.

MAIN TRANSMISSION COUNTERSHAFT (LAYSHAFT) AND GEARS

Models With Twelve-Speed Shuttle Transmission

233. The upper (mainshaft) must be removed as outlined in paragraph 232 and pto lower shaft (46—Fig. 243) and drive gear (45) should be removed as outlined in paragraph 230, before removing countershaft. A $7/16$ inch UNF screw approximately 75-100 mm (3-4 inches) long can be threaded into front of pto lower shaft and used to assist in removal of lower pto shaft.

Remove snap ring (25—Fig. 244) and spacer (30) from rear of countershaft (31), bump shaft forward enough to remove split ring (29) from groove. Remove snap ring (35) and thrust washer (84), then push countershaft (31) to the rear as far as possible. Slide constant mesh gear (34) forward as far as possible, lift snap ring (86) from groove and slide snap ring forward. Push countershaft (31) toward rear and remove reverse gear (34) with synchronizer cone (74). Remove snap ring (86) and synchronizer ring (75), then remove synchronizer assembly (76 through 79) as a unit. Slide synchronizer ring (80), synchronizer cone (85), gear (33) and washer (88) from shaft. Push countershaft (31), bearing (32), snap ring (87) and bearing (32) out toward front, through center web, then remove gears (27 and 28).

Synchronizer parts (76 through 79) should be removed as an assembly and separated only if necessary. Wrap synchronizer assembly with a cloth to catch poppet parts (77 and 78) when coupling (76) is removed from hub (79). Clean, inspect and renew any components that are damaged or worn excessively. Place synchronizer cones (74 and 85) and synchronizer rings (75 and 80) on respective gears (33 and 34). Make sure that each ring is seated squarely, then measure clearance between ring and gear at several places around gear, using a feeler gauge. Install new synchronizer ring if clearance is less than 0.5 mm (0.020 inch). This clearance can be checked with gearbox assembled, but minimum clearance is increased to 0.8 mm (0.030 inch).

Fig. 245—Synchronizer rings for unit used on countershaft are different than similar parts used on mainshaft. Holes (H) are used when checking and adjusting position of shift forks and must be aligned when assembling synchronizer. Refer to Fig. 244 for legend.

Fig. 246—Cross section of twelve-speed shuttle transmission.

1. Range selector shaft (Fig. 231)
8. Pto input shaft (Fig. 242)
11. Transmission input shaft (Fig. 242)
13. Second gear (Fig. 244)
16. High range synchronizer cone (Fig. 234)
20. First gear (Fig. 244)
21. Fourth gear (Fig. 244)
22. Front-wheel-drive gear (Fig. 234)
24. Third gear (Fig. 244)
27. Third gear (Fig. 244)
28. Fourth gear (Fig. 244)
31. Medium/Low range shift collar (Fig. 234)
33. Forward gear (Fig. 244)
34. Reverse gear (Fig. 244)
35. Low range gear (Fig. 234)
37. Upper range shaft (Fig. 234)
45. Pto lower gear (Fig. 243)
46. Pto lower shaft (Fig. 243)
51. Lower range shaft (Fig. 234)
56. Middle range gear (Fig. 234)
58. Driven gear (Fig. 234)
64. Front-wheel-drive gear (Fig. 234)
76. Shifting coupler (Fig. 244)
81. Forward gear (Fig. 242)
82. Reverse gear (Fig. 242)
91. Shifting coupling (Fig. 244)

When reassembling the synchronizer shift components, align centralizing holes in shift collar (76) and hub (79). Synchronizer rings (75 and 80) on countershaft are different than similar parts on mainshaft, because countershaft rotates in opposite direction. Special tools (MF 415 and MS 550) are available to assist holding the three poppets (77 and 78) into hub (79) while installing coupling (76).

Reassemble by reversing removal procedure. Always install new snap rings when reassembling. Position snap ring (29) around shaft in the unsplined area and install front bearing (32) on shaft before installing shaft in housing. Refer to Fig. 246 for cross section of assembled transmission. Refer to paragraph 232 for assembly of mainshaft and selection of thrust washer (49—Fig. 244).

Coat gasket (43—Fig. 243) lightly with "Hylomar" or equivalent sealer before installing. Coat threads of screws attaching cover (36) and screws attaching retainer housing (3—Fig. 242) lightly with "Hylomar"

or equivalent sealer before tightening to 54-61 N·m (40-45 ft.-lbs.) torque.

REVERSE IDLER GEAR AND SHAFT

Models With Twelve-Speed Shuttle Transmission

234. To remove reverse idler gear (54—Fig. 244), shaft (57) and related parts, first remove the transmission as outlined in paragraph 220. Remove the top cover and shift levers as outlined in paragraph 221 and the shift rails and forks as outlined in paragraph 224. Refer to paragraph 229 and remove the pto lower shaft front bearing and retainer, then refer to paragraph 231 and remove the transmission input shaft.

Carefully withdraw reverse idler shaft (57) from front and lift idler gear (54), spacer (52), washers (51) thrust washers (50) and bearing rollers (53) through top opening, being careful not to drop any parts into housing. Each bearing path (53) contains 23 rollers.

Some models have only a single gear at (54) and a spacer is installed.

Assemble needle rollers (53) and washers (51) in reverse idler gear using petroleum jelly. Position thrust washers (50), spacer (52) and gear and bearing assembly in housing, then install idler shaft (57), using new "O" ring (56). Complete reassembly by reversing disassembly procedure.

MAIN DRIVE BEVEL DRIVE GEARS AND DIFFERENTIAL

The main drive bevel pinion and ring gear reduction ratio is different for different models. Model 340 tractors are equipped with direct drive axles, while other models are equipped with a planetary reduction at axle outer end.

DIFFERENTIAL AND BEVEL RING GEAR

All Models

235. R&R AND OVERHAUL. The ring gear and differential unit can be removed from the differential housing after removing the rear axle housing assembly from the left side as outlined in paragraph 242, 244 or 246.

To disassemble the removed differential unit, first scribe alignment marks on both halves of differential case to facilitate correct reassembly. Pull cone of bearing (4—Fig. 247) from differential lock coupler cap (8), then remove cap screws (7). Remove coupler cap and separate differential case halves (9 and 16). Differential pinions (13), side gears (11) and cross (12) can then be removed.

The main drive bevel ring gear (15) is secured to differential case half with either special bolts and

Fig. 247—Exploded view of differential and bevel gears typical of most models. Differential lock mechanism (5 and 35) is different and carrier housings (1 and 2) are not used on direct drive 340 tractors. Other differences may be noted depending upon pto drive.

1. Carrier housing (Right)	14. Thrust washer
2. "O" ring	15. Bevel ring gear
3. Spacer shield	16. Differential case half
4. Bearing cup & cone	17. Bearing cup
5. Differential lock fork	19. Carrier housing (Left)
6. Jaw coupler	20. Snap ring
8. Coupler cap	21. Bearing
9. Differential case half	22. Bevel pinion gear
10. Thrust washer	23. Bearings
11. Side cover	24. Locating pin
12. Cross	25. Retainer housing
13. Pinion gear	26. Bearing
	27. Spacer
	28. Adjusting nut
	29. Locking pins
	30. Pto drive gear
	31. Bearing
	32. Sleeve & keeper
	33. Thrust washer
	34. Thrust washer
	35. Differential lock shaft

Fig. 248—On 340 models, the oil tray (O) should be in place as shown to direct lubricant.

Fig. 249—View showing special tool MF245D, a straight edge (S) and feeler gauge (F) to determine correct thickness of spacer shield to install.

nuts (18) or with rivets. Note that the ring gear is a press fit on case and may also be epoxy bonded to case if previously serviced.

Inspect all parts for scoring, chipping, wear or other damage and renew if necessary. Backlash between pinion gears (13) and side gears (11) should be 0.08-0.28 mm (0.003-0.011 inch). It is suggested that cross (12) and all pinion gears (13) be renewed together.

To reassemble differential, reverse the disassembly procedure while noting the following special instructions. If bevel ring gear (15) was removed, install using special epoxy bonding kit. Be sure to follow kit instructions and cure epoxy for the correct time at the specified temperature. Apply "Loctite 270" or equivalent to threads of ring gear retaining screws, then tighten retaining nuts (18) evenly and to 160 N·m (120 ft.-lbs.) torque. Be sure to align previously affixed scribe marks on differential case and tighten retaining cap screws (7) to 108 N·m (80 ft.-lbs.) torque.

To reinstall differential, reverse the removal procedure. On 340 models with direct drive axles, make sure that oil tray is attached as shown in Fig. 248. Refer to paragraph 236 for all models if differential case halves and/or carrier bearings were renewed.

236. CARRIER BEARING PRELOAD. Differential carrier bearing preload is adjusted by installing the correct thickness spacer shield (3—Fig. 247) between bore in carrier housing (1) and cup for the right side bearing (4). To check preload adjustment using

special tool (MF 245D or equivalent) refer to Fig. 249. Assemble differential assembly and left carrier housing (19—Fig. 247) in housing. Remove cup for right bearing (4) and the spacer shield (3) from the right carrier housing (1). Position the cup for bearing (4) over the bearing cone, then install bar (B—Fig. 249) of special tool over the bearing cup and cone as shown. Attach bar (B) to housing with two screws tightened to 27 N·m (20 ft.-lbs.) torque and install the two special blocks (T). Locate a straightedge (S) across blocks and measure gap between center of bar and straightedge with a feeler gauge as shown at (F). Install spacer shield (3—Fig. 247) equal to the clearance measured by feeler gauge (F—Fig. 249). Spacer shields may be marked for identification by a series of punch marks or dots to indicate thickness as follows:

Eight Marks	0.45-0.53 mm (0.019-0.021 inch)
Seven Marks	0.61-0.66 mm (0.024-0.026 inch)
NO Marks	0.74-0.79 mm (0.029-0.031 inch)
One Mark	0.86-0.91 mm (0.034-0.036 inch)
Two Marks	0.99-1.04 mm (0.039-0.041 inch)
Three Marks	1.12-1.17 mm (0.044-0.046 inch)
Four Marks	1.25-1.30 mm (0.049-0.051 inch)
Five Marks	1.37-1.42 mm (0.054-0.056 inch)

Install the spacer shield with curved section as shown in Fig. 250. Remainder of reassembly is reverse of disassembly procedure.

Fig. 250—Cross section of spacer shield (3) correctly installed. Refer to Fig. 247 for exploded view.

DIFFERENTIAL LOCK

All Models

237. OPERATION. The mechanically actuated differential lock assembly is standard equipment on all models. When the differential lock foot pedal is depressed, the axle half of coupler (6—Fig. 247) is forced inward to contact the differential case half of coupler cap (8). If slippage is occurring at either wheel, depressing the lock pedal will lock the differential case to the right axle. The differential and both rear drive wheels will rotate together as a unit. Pressure between the coupler dogs should keep the differential lock engaged and foot pedal may be released. When ground traction on both rear wheels becomes equal again, coupler dog contact pressure will be relieved and the coupler will automatically disengage.

238. ADJUSTMENT. Two different operating mechanisms are used and different adjustment procedures are required. Refer to Fig. 254 for exploded view of the linkage used for direct drive axles of 340 models. Refer to Fig. 255 for linkage typical of models with other models with planetary reduction axles.

To check adjustment of differential lock linkage on Model 340 tractors, depress pedal and measure distance between pedal and cab floor or step plate. Correct clearance is 6-12 mm (¼-½ inch). If lock does not engage or if clearance between pedal and floor is

Fig. 251—Differential lock linkage on Model 340 tractors can be adjusted by loosening clamp screw (Arrow) and relocating operating lever.

Fig. 252—External view of differential lock linkage typical of models except 340 with direct drive axles. Adjustment screw is shown at (12). Refer to Fig. 255 for legend.

incorrect, loosen clamp screw (17—Fig. 251), reposition pedal, then tighten clamp screw to 115 N·m (85 ft.-lbs.) torque.

On all except direct drive axle models, check operation of differential lock linkage. Make sure that return spring is installed, linkage parts are clean and spring in operating plunger is free to move. When disengaged, clearance at (12—Fig. 252), between operating lever and adjusting screw, should be 1.0 mm (0.040 inch). If incorrect, lift rubber boot, loosen lock nut, then turn adjusting screw until clearance is correct. Recheck operation, then tighten lock nut and

Fig. 253—Cross section of differential lock typical of models with planetary reduction and disc brakes.

reposition rubber boot if clearance is correct. Refer to Fig. 253 for cross section.

On all models, refer to paragraph 239 if operating mechanism is rusted, dirty or for any other reason cannot be adjusted properly.

239. R&R AND OVERHAUL. To remove the differential lock, first drain transmission housing. Disconnect the brake and differential lock linkage, then refer to paragraph 242 for removing rear axle from right side of early 340 models with drum brakes. Refer to paragraph 244 for removing direct drive rear axle housing from right side of late 340 models with disc brakes. Refer to paragraph 246 for removing rear axle housing from right side of models with planetary reduction. Operating mechanism (Fig. 254) of model 340 tractors is located in the right axle housing. The

Fig. 254—Exploded view of differential lock mechanism used on direct drive 340 models. Refer to Fig. 255 for other models.

1. Shaft
2. Fork bracket
3. Shoe (2 used)
4. Actuating fork
5. Pivot bolt (2 used)
6. Oil seal
7. Lock coupling
10. Pedal bracket
11. Actuating cam
12. Pin
13. Bushing
14. Bracket
15. Return spring
16. Pedal
17. Clamp bolt

operating shaft (5—Fig. 255) of other models passes through right axle housing and the differential carrier housing (1). Clean dirt and rust from shafts (1—Fig. 254 or 5—Fig. 255) before attempting to remove.

If renewal is necessary, bushings for shafts (1—Fig. 254 or 5—Fig. 255) should be removed and installed using piloted drivers of the proper size. Make sure that plunger pin (24—Fig. 255) is free to operate in bore, before assembling. Remainder of assembly is reverse of disassembly. Refer to paragraph 238 for adjustment procedure.

MAIN DRIVE BEVEL GEARS

All Models

240. REMOVE AND REINSTALL. The main drive bevel ring gear (15—Fig. 247) and bevel pinion gear (22) must be renewed as a matched set. Different reduction ratios are used in various applications. Model 340 tractors are not equipped with planetary final reduction, and thus the bevel gear reduction is the final reduction. Normal installed reduction ratio of the bevel gears is as follows:

All Model 340
Ratio 6.167:1
Teeth . . . . 6×37
Standard Axle Models 350 and 360
Ratio. 4.375:1
Teeth . . . . 8×35
Standard Axle Model 399
Ratio. 3.889:1
Teeth . . . . 9×35
Optional (40 km/hr) Axle Model 399
Ratio. 3.454:1
Teeth . . . . 11×38
All Models With 8 or 12 Speed Shuttle Gearbox
Ratio. 3.454:1
Teeth . . . . 11×38

To remove the bevel ring gear, first remove the differential assembly as outlined in paragraph 235. The main drive bevel ring gear is secured to differential case half with either special bolts and nuts or with rivets. Rivets can be drilled out using a 13 mm (½ inch) drill after center punching rivet head. Retaining nuts are installed with thread locking compound and removal is easier if nuts are heated. The ring gear is a press fit on case and may also be epoxy bonded to case.

To remove the bevel pinion gear, first remove the differential as outlined in paragraph 235 and hydraulic lift cover as outlined in paragraph 285. Removal procedure is different depending upon pto and transmission type. Refer to the following procedures.

On models with independent pto, split tractor between differential housing and transmission housing

Fig. 255—Exploded view of differential lock mechanism typical of all models except 340. Hydraulic pressure may be directed against shaft (5) of some models to move shaft and fork (3). Differential carrier housing (1) is attached to right side axle housing with screws (S).

1. Carrier housing	7. Collar	13. Control lever boot	19. Screws
2. Lock coupling	8. Snap ring	14. Return spring	20. Pin
3. Actuating fork	9. "O" ring	15. Control lever	21. Housing
4. Pin	10. Locknut	16. Bar	22. Spring
5. Shaft	11. Boot	17. Block	23. Plunger pin
6. Spring	12. Adjusting screw	18. Pivot	24. Operating pin

as outlined in the appropriate paragraph in the transmission section. Remove pto output shaft and cover from left side of rear axle center housing. Disconnect hydraulic lines and remove pump support pins (S—Fig. 256A), then remove hydraulic pump and independent pto clutch unit through front of differential housing.

It is not necessary to split models not equipped with independent pto. It is, however, necessary to remove the coupler tube and pinion drive shaft. Remove side cover from left side of housing and move pto driven gear forward on pto shaft splines. Remove ground speed pto drive gear from pinion shaft.

On all models, remove screws attaching retainer (25—Fig. 247) to housing, then thread two of the screws into tapped, jackscrew holes in retainer. Tighten the two jackscrews evenly to force retainer and pinion assembly from housing bore.

Refer to paragraph 241 for overhauling the pinion and bearings assembly and to paragraph 235 for installing new ring gear. Refer to paragraph 236 to check and adjust differential carrier bearing preload, before installing the bevel pinion assembly.

Make sure that locating pin (24—Fig. 256) is in place and correctly aligned with hole in housing be-

fore pressing retainer (25) into place. Tighten screws attaching retainer (25) to 108 N·m (80 ft.-lbs.) torque. Complete assembly by reversing removal procedure. Always install new split pins, "O" rings and gaskets. Carefully locate the pump on ipto clutch and install supports (S—Fig. 256A). Tighten nuts or screws retaining support (S) to right side, but leave left side support retaining nuts or screws loose until tractor assembly is complete. Be sure that engine will not start and crank engine with starter, then tighten nuts or screws retaining support to left side of axle center housing. This procedure will ensure correct alignment of shafts.

241. OVERHAUL. Refer to paragraph 235 for renewal of bevel ring gear and for overhaul of the removed differential assembly. Refer to Fig. 256 for illustrations of typical bevel pinions and associated parts. Parts shown at top are typical of single speed pto, parts shown in middle view are typical of models with shiftable live or ground speed pto and parts shown in lower view are typical of two speed (540-1000 rpm) shiftable pto. Disassembly and reassembly procedures are similar for all models, even though

Fig. 256—Exploded view of typical bevel pinion gears and related parts. Parts shown at top are typical of single speed pto, parts shown in middle view are typical of models with shiftable live or ground speed pto and parts shown in lower view are typical of two speed (540-1000 rpm) shiftable pto. Sleeve (32A) for use with standard hydraulic pump is longer than similar sleeve for use with high volume pump.

20. Snap ring
21. Bearing
22. Bevel pinion gear
23. Bearings
24. Locating pin
25. Retainer housing
26. Bearing
27. Spacer
28. Adjusting nut
29. Locking pins (2 used)
30. Pto drive gear
31. Bearings
32. Sleeve & keeper
33. Thrust washer
34. Thrust washer

Fig. 256A—Hydraulic pump supports (S) are attached to each side of rear axle center housing.

parts may be different. Refer to paragraph 240 for removal of pinion assembly.

Nut (28) is locked in place on all models by two locking pins (29) driven between shaft splines and threads of nut. To remove, it is necessary to destroy nut (28) by cutting. Make sure that new nut is available before removing. Clamp nut (28) in a soft jawed

vise with flats of nut adjacent to locking pins (29) against jaws. Use a chisel to cut nut about ⅔ through at two places 90° from locking pins using a chisel. Nut should fracture during the second cut and make removal easy. Do not attempt to reuse nut. Bump pinion (22) from bearings after removing nut.

Examine all parts for wear or other damage and renew parts as necessary. Even though similar parts are identified by the same number in Fig. 247, parts may be different and the required parts may have different factory part numbers.

When reassembling, adjust preload of bearings (23 and 26) by tightening nut (28). Correct bearing preload will require 206-245 N·m (18-22 in.-lbs.) rolling torque to rotate pinion shaft in bearings. Bearing preload can be checked with a spring scale attached to string wrapped around pinion shaft and bearing retainer clamped in a vise. If rolling torque is checked with a spring scale and string, force should be 10-11 kg (21-25 lbs.). Lock adjustment by driving lock pins (29) into threads of nut in two adjacent splines under one of the flats. Drive locking pins in until flush with front of locking nut. Recheck rolling torque after locking nut and tapping shaft to centralize bearings.

Lubricate pinion assembly thoroughly, align pin (24) with hole in housing, then tighten retaining nuts to 108 N·m (80 ft.-lbs.) torque.

REAR AXLE AND FINAL DRIVE

All tractors except early 340 models are equipped with a disc type brake located at the inner end of each axle. All tractors except 340 models are equipped with a planetary type reduction unit located in outer end of each rear axle housing. All Model 340 tractors are equipped with direct drive rear axle. Early 340 tractors are equipped with drum and shoe type brakes located at the outer ends; later 340 tractors are fitted with disc type brakes located at inner ends of axle shafts. Refer to the appropriate following paragraphs for removal and overhaul procedures.

REAR AXLE ASSEMBLY

Early 340 Models With Drum Brakes

242. REMOVE AND REINSTALL. Axle shafts can be removed without removing axle housings as outlined in paragraph 243.

To remove the complete axle shaft and housing (29—Fig. 257) as a unit, first drain oil from differential housing, raise and support rear of tractor, then remove the rear wheel. Release parking brake and detach cable from actuating cam shaft. Disconnect

Fig. 257—Exploded view of rear axle, housing and brake drum typical of type used on early 340 models.

20. Brake drum
21. Axle shaft
22. Oil seal retainer
    (outer)
23. Retainer
24. Bearing cup
25. Bearing cone
26. Retaining collar

27. Shims
28. Oil seal (inner)
29. Axle housing (left)
30. Bushing
31. Plug
32. Gasket
33. Bearing cup

hydraulic brake lines and immediately cover all openings to prevent the entrance of dirt. If right axle housing is being removed, remove the differential lock pedal. Unbolt and remove stabilizer bracket from outer end of axle housing and detach lower lift link from housing inner flange. Detach lift link from upper knuckle and remove the lower lift link. Remove the fender assembly or support the cab so that axle housing can be removed, then support the axle housing so that it can be safely separated from the tractor. Remove screws and nuts securing axle housing to the center housing, then carefully move the axle assembly away from center housing.

> CAUTION: Be careful if axle housing is removed from left side, because differential may fall from center housing. Remove differential unit as axle housing is removed to prevent falling and subsequent damage.

Coat housing bore with "Loctite 545" or equivalent before driving seal (28) into position with lip toward differential. If removed, reattach oil trough to left axle housing. If removed, install lower lift arm pin, tightening internal retaining nut to 149-163 N·m (110-163 ft.-lbs.) torque. If both axle assemblies are removed, install differential and left side axle housing before installing housing on right side.

Stick new gasket to center housing using petroleum jelly before positioning axle housing against center housing. Carefully align axle shaft splines with splines of differential side gears and holes in axle housing with attaching studs. Tighten retaining stud nuts to 115 N·m (85 ft.-lbs.) torque. If necessary, refer to paragraph 236 to check and adjust differential carrier bearing preload. Remainder of assembly is the reverse of disassembly procedure. Coat threads of fender or cab stabilizer bracket with "Loctite 270" or equivalent, then tighten nuts to 230 N·m (170 ft.-lbs.) torque. Tighten wheel retaining nuts to 270 N·m (200 ft.-lbs.) torque. Fill transmission housing to correct level with oil and bleed air from brake lines.

243. R&R AXLE SHAFT. The axle shafts can be removed without removing axle housings if desired. Raise and support rear of tractor, then remove rear wheel and tire. Engage the differential lock if right axle is removed. Remove drum brake retaining screws and pull drum (20—Fig. 257) from axle flange. Remove cap screws attaching retainer (23) to axle housing, then withdraw axle shaft (21) with bearing assembly and shims (27) from housing. It may be necessary to remove the complete axle housing to reinstall drive axle if differential lock coupling does not remain in alignment while drive axle is removed.

To reinstall, reverse the removal procedure while noting the following special instructions: Install inner oil seal (28) with lip toward inside. Grease lip of seal before inserting axle shaft. To check axle shaft end play, first make sure that opposite axle is in place and moved fully outward. Install axle assembly and tighten screws attaching retainer (23) to 74 N·m (55 ft.-lbs.) torque, then mount a dial indicator so that axle end play can be measured. Correct axle end play is 0.05-0.20 mm (0.002-0.008 inch) and can be changed by adding or removing shims (27), which are available in various thicknesses. Note that end play is between the two axle shafts and is controlled by the total thickness of shims installed at both ends. Normally, the right side is assembled with one 0.38 mm (0.015 inch) and one 0.13 mm (0.005 inch) shim. The left side can be initially assembled with two 0.51 mm (0.020 inch), one 0.38 mm (0.015 inch) and two 0.13 mm (0.005 inch) shims. It is usually necessary to change only the thickness of shims on left side to obtain correct end play.

340 Models With Disc Brakes

244. REMOVE AND REINSTALL. Axle shafts can be removed without removing axle housings as outlined in paragraph 245.

To remove the complete axle shaft and housing as a unit, first drain oil from transmission housing, raise and support rear of tractor, then remove the rear wheel. Release parking brake, then detach cable and return spring. Disconnect hydraulic brake lines and immediately cover all openings to prevent the entrance of dirt. If left axle housing is to be removed, remove necessary lines and the auxiliary hydraulic manifold. If right axle housing is being removed, remove linkage for differential lock that would interfere with removal and disconnect lock switch wiring. Unbolt fender or cab stabilizer bracket from outer end of axle housing and detach lower lift link from housing inner flange. Detach lift link from upper knuckle and remove the lower lift link. Remove the fender assembly or support the cab so that axle housing can be removed, then support the axle housing so that it can be separated from the tractor.

> **CAUTION: The axle housing assembly is heavy and awkward to handle. Be careful to support and move it safely. If axle housing is removed from left side, differential may fall from center housing. Remove differential unit as axle housing is removed to prevent falling and subsequent damage.**

Remove screws and nuts securing axle housing to the center housing, then carefully move the axle assembly away from center housing. Remove the "O" ring from carrier flange plate. If both axle assemblies are removed, install differential and left side axle housing before installing housing on right side.

Stick new "O" ring to center housing using petroleum jelly before positioning axle housing against center housing. Carefully align axle shaft splines with splines of differential side gears and holes in axle housing with attaching studs. Tighten retaining screws and stud nuts to 104-156 N·m (77-115 ft.-lbs.) torque. If necessary, refer to paragraph 236 to check and adjust differential carrier bearing preload. Remainder of assembly is the reverse of disassembly procedure. Coat threads of fender or cab stabilizer bracket with "Loctite 270" or equivalent, then tighten nuts to 230 N·m (170 ft.-lbs.) torque. Tighten wheel retaining nuts to 325 N·m (240 ft.-lbs.) torque. Fill transmission housing to correct level with oil and bleed air from brake lines.

245. R&R AXLE SHAFT. The axle shafts (1—Fig. 258) can be removed without removing axle housings if desired. Lock parking brake ON and make sure that brake stays ON until assembly is completed to hold brake discs from falling. Engage the differential lock if right axle is removed. Raise and support rear of tractor, then remove rear wheel and tire. Remove cap screws attaching outer retainer (6) to axle housing (5), then withdraw axle shaft (1) with bearing assembly (9) and shims (8) from housing, differential side gear, differential lock coupling and brake discs. It may be necessary to remove the complete axle housing to reinstall drive axle if parts do not remain in alignment while drive axle is removed.

To disassemble, bearing retaining collar (10) must be drilled, then fractured with a chisel. Bearing (9) can be pressed from axle and new oil seal (4) can be installed.

Fig. 258—Partially exploded view of direct drive axle used on late 340 tractors. Brakes are located at inner end of axle.

1. Wheel axle
2. Dust shield
3. Wheel stud
4. Seal
5. Axle housing
6. Drive cover
7. Seal ring
8. Shims
9. Bearing cup and cone
10. Retainer ring

To assemble, coat retainer housing bore (6) for outer seal (4) with "Loctite 545" or equivalent, then press new seal in place with lip toward inside. Press new bearing cup into retainer housing if required, then fill space between oil seal (4) and bearing (9) with high melting point grease. Position retainer housing over axle shaft being careful not to damage seal, then press bearing cone onto axle until seated. Heat new retainer (10) until dull red, then press against inner bearing race before collar cools. After cooling, fill bearing cavity with approximately 85 g (3 oz.) of high melting point grease. Lip of inner oil seal should face in toward differential. If renewed, coat outer surface of seal with "Loctite 545" or equivalent, before seating in bore of axle housing.

To reinstall, reverse the removal procedure while noting the following special instructions: Stick new "O" ring to axle housing using petroleum jelly. To check axle shaft end play, first make sure that opposite axle is in place and moved fully outward. Install axle assembly with more shims than necessary and at least three equally spaced screws tightened to 74 N•m (55 ft.-lbs.) torque. Mount a dial indicator so that axle end play can be measured. Correct axle end play is 0.05-0.20 mm (0.002-0.008 inch) and can be changed by adding or removing shims (8), which are available in various thicknesses. Note that end play is between the two axle shafts and is controlled by the total thickness of shims installed at both ends. It is usually necessary to change the thickness of shims on only one side to obtain correct end play.

All Except 340 Models

246. REMOVE AND REINSTALL. Wheel axle shafts and planetary units can be removed without removing axle housings as outlined in paragraph 247.

To remove the complete axle shaft and housing as a unit, first drain oil from transmission housing, raise and support rear of tractor, then remove the rear wheel. Release parking brake, then detach cable and return spring. Disconnect hydraulic brake lines and immediately cover all openings to prevent the entrance of dirt. If left axle housing is to be removed, remove necessary lines and the auxiliary hydraulic manifold. If right axle housing is being removed, remove linkage for differential lock that would interfere with removal and disconnect lock switch wiring. Unbolt fender or cab stabilizer bracket from outer end of axle housing and detach lower lift link from housing inner flange. Detach lift link from upper knuckle and remove the lower lift link. Remove the fender assembly or support the cab so that axle housing can be removed, then support the axle housing so that it can be separated from the tractor.

CAUTION: The axle housing assembly is heavy and awkward to handle. Be careful to support and move it safely. If axle housing is removed from left side, differential may fall from center housing. Support the differential unit as axle housing is removed to prevent it from falling.

Remove screws and nuts securing axle housing to the center housing, then carefully move the axle assembly away from center housing. Remove the "O" ring from carrier flange plate. If both axle assemblies are removed, install differential and left side axle housing before installing housing on right side.

Stick new "O" ring to center housing using petroleum jelly before positioning axle housing against center housing. Carefully align axle shaft splines with splines of differential side gears and holes in axle housing with attaching studs. Tighten retaining screws and stud nuts to 104-156 N•m (77-115 ft.-lbs.) torque. If necessary, refer to paragraph 236 to check and adjust differential carrier bearing preload. Remainder of assembly is the reverse of disassembly procedure. Coat threads of fender or cab stabilizer bracket with "Loctite 270" or equivalent, then tighten nuts to 230 N•m (170 ft.-lbs.) torque. Tighten wheel retaining nuts to 325 N•m (240 ft.-lbs.) torque. Fill transmission housing to correct level with oil and bleed air from brake lines.

PLANETARY ASSEMBLY

All Models With Planetary Final Drive

247. REMOVE AND REINSTALL. To remove the planetary assembly, first drain oil from planetary housing, raise and support rear of tractor, then remove the rear wheel. Lock parking brake ON and make sure that brake stays ON until assembly is completed to hold brake discs from falling. Scribe a mark across axle flange, ring gear (20—Fig. 259 or Fig. 260) and outer drive cover (6) to facilitate alignment when assembling, then remove nuts from screws holding planetary together. Lift outer housing, planetary assembly and ring gear away from axle housing, being careful not to withdraw inner axle from differential side gears and brakes.

Reinstall by reversing removal procedure. All sealing surfaces should be cleaned and inspected for damage that could cause leakage. Always install new gaskets on sides of ring gear. Tighten nuts securing outer housing and ring gear to 90-120 N•m (66-89 ft.-lbs.) torque on models with heavy duty axle. Tighten nuts securing outer housing and ring gear to 68-74 N•m (50-55 ft.-lbs.) torque on models with normal duty axles. The heavy duty axle with 4.8:1 reduction ratio is used in 399 models.

248. BEARING ADJUSTMENT. Planetary carrier bearing preload is adjusted by installing shims (25—Fig. 259 or Fig. 260) of different thicknesses under

Fig. 259—Exploded view of normal duty wheel axle and planetary assembly used on 350, 355 and 360 models. Heavy duty planetary, used on dome models, is shown in Fig. 260.

1. Wheel axle	14. Thrust washer
2. Dust shield	15. Needle rollers
3. Wheel stud	16. Spacer washer
4. Seal	17. Pinion gear
5. Plug	18. Sun gear
6. Drive cover	19. Bushing
7. Gasket (same as 21)	20. Ring gear
8. Bearing cup	21. Gasket (same as 7)
9. Bearing cone	22. Drive axle
10. Thrust washer halves	23. Bearing cone
11. Planetary carrier	24. Bearing cup
12. Roll pin	25. Shim
13. Pinion shaft	

Fig. 260—Exploded view of heavy duty wheel axle and planetary assembly used on 399 models.

1. Wheel axle	14. Thrust washer
2. Dust shield	15. Needle rollers
3. Wheel stud	16. Spacer washer
4. Seal	17. Pinion gear
5. Plug	18. Sun gear
6. Drive cover	20. Ring gear
7. Gasket (same as 21)	21. Gasket (same as 7)
8. Bearing cup	22. Drive axle
9. Bearing cone	23. Bearing cone
10. Snap ring	24. Bearing cup
11. Planetary carrier	25. Shim
12. Retaining pins	26. Seal
13. Pinion shaft	

the bearing cup (24). Special tools (MF 267A, MF 267-1 and MF 267-2) are available for checking bearing preload and, if available, should be used as follows:

Remove the wheel axle, planetary outer housing, planetary unit and ring gear as outlined in paragraph 247. Bolt the ring gear (20) to the outer housing (6), using four of the original bolts spaced evenly around circumference and four of the wheel retaining nuts as spacers. Apply the handbrake fully and engage the differential lock (if right axle is being removed), then withdraw the drive axle (22) from the differential side gear, differential lock coupling and brake discs. It may be necessary to remove the complete axle housing to reinstall drive axle if parts do not remain in alignment while drive axle is removed. Use a suitable

puller to remove bearing cup (24) and any shims (25) that were previously installed. Use special sleeve (MF 267-1 for normal duty axle, MF 267-2 for heavy duty axle), position the removed bearing cup on the sleeve and install the assembled wheel axle and planet assembly into the bearing and special tool. Measure clearance between the ring gear and the two lugs of special tool. Clearance should be the same for each of the two lugs. Subtract the clearance measured from 1.0 mm (0.04 inch) and install shims (25) equal to this thickness in axle housing before installing bearing cone (24). Shims are available in thicknesses of 0.13, 0.25 and 0.38 mm (0.005, 0.010 and 0.015 inch). Refer to paragraph 247 for remainder of assembly.

If special tool is not available, remove the wheel axle, planetary outer housing, planetary unit and ring gear as outlined in paragraph 247. Apply the handbrake fully and engage the differential lock (if right axle is being removed), then withdraw the drive axle (22) from the differential side gear, differential lock coupling and brake discs. It may be necessary to remove the complete axle housing to reinstall drive axle if parts do not remain in alignment while drive axle is removed, but housing should remain on tractor while checking bearing preload. Use a suitable puller to remove bearing cup (24) and any shims (25) that were previously installed. Add approximately 0.76 mm (0.030 inch) thickness to the shims previously installed and reinstall bearing cup (24). Do not install drive axle (22), but install planetary assembly and ring gear using only three equally spaced retaining bolts finger tight. Bump outer end of wheel axle (1) sharply to make sure that bearings are seated. Equalize gap between axle housing and ring gear by tightening the three retaining bolts as required. When gap is equal, measure gap at three places next to the installed bolts and record the measurements. Remove the three retaining bolts, wheel axle, ring gear and planetary assembly. Remove the bearing cup (24) and remove shims equal to the gap plus 0-0.025 mm (0-0.0010 inch) to establish proper bearing preload. Reinstall drive axle and refer to paragraph 247 for remainder of assembly.

Normal Duty Axle and Planetary Unit

The 4.8:1, heavy duty planetary reduction unit installed on 399 models is serviced using procedures described in paragraph 250. For servicing the normal duty unit with 3.75:1 installed on 350, 355 and 360 models, refer to the following paragraph 249.

249. OVERHAUL. For servicing the normal duty unit with 3.75:1 reduction ratio, first refer to paragraph 247 and remove the planetary unit, then lift off ring gear (20—Fig. 259). Drive out pinion shaft retaining pins (12), screw a ⅜ inch cap screw into threaded end of pinion shaft (13) and pull shafts from carrier. Carefully withdraw the three pinion gears (17) with thrust washers (14), loose rollers (15) and spacer washers (16). Each planetary pinion gear contains two rows of loose needle rollers (15) separated by a washer (16). Each path of rollers (15) contains twenty-nine individual rollers.

> **CAUTION: Do not lose, damage or mix rollers from other gears and make sure to account for all rollers, making sure that none stick to gear teeth or other surface.**

Slide sun gear (18) out through largest opening in planetary carrier (11). Drive bushing (19) into plane-

tary carrier, then remove bushing. Remove inner bearing cone from planetary hub using an appropriate bearing puller, then use special tool (MF 265A or equivalent) and a 50 ton press to remove planet gear hub from axle splines. Remove both halves of thrust washer (10), then press wheel axle out of drive cover (6).

Clean, inspect and renew any components that are damaged or worn excessively. Petroleum jelly should be used to lubricate and hold components, including needle rollers, in position while assembling.

Dust shield (2) is retained by wheel studs and should not be removed unless renewal is necessary. Apply a light coat of "Hylomar" or similar sealer to outer surface of wheel axle seal (4), then install seal in drive cover (6) with metal side out and protruding 2 mm (0.080 inch) above flush with cover bore. Coat seal lip and fill cavity between seal and bearing cup (8) with petroleum jelly. Assemble wheel axle (1), drive cover (6) and bearing (8 and 9), pressing bearing cone (9) on shaft until fully seated.

Install thrust washer halves (10) and measure clearance between thrust washer halves and edge of shaft groove using a feeler gauge. Desired clearance is 0-0.025 mm (0-0.001 inch) and is adjusted by changing thickness of thrust washer halves which are available in thicknesses of 5.84-5.89, 5.90-5.94, 5.95-5.99, 6.00-6.04, 6.04-6.09, 9.10-6.14, 6.14-6.20, 6.20-6.25, 6.25-6.30 mm (0.230-0.232, 0.232-0.234, 0.234-0.236, 0.236-0.238, 0.238-0.240, 0.240-0.242, 0.242-0.244, 0.244-0.246 and 0.246-0.248 inch). Make sure that both halves are the same thickness and that thrust washer halves are fully seated in groove, then press the planet carrier (11) on wheel axle shaft until it bottoms.

Bushing (19) is presized and should not require machining after installation. Install bushing flush with chamfered edge in planet carrier bore. Preassemble planet gears (17), bearing rollers (15), spacer washer (16) and thrust washers (14), using petroleum jelly. Twenty-nine bearing rollers are in each path and each planet gear has two paths of bearings separated a by washer (16). Install sun gear (18) and the preassembled pinion gears in carrier (11). Larger thrust washers with squared ends are installed in the sun gear removal and installation opening. Insert pinion shafts (13) in carrier, aligning shaft retaining pin holes with corresponding holes in carrier, then install pins (12).

Install wheel axle and planetary assembly and adjust planetary carrier bearing preload as outlined in paragraphs 247 and 248.

Heavy Duty Axle and Planetary Unit

The normal duty planetary reduction unit used on 350, 355 and 360 models is serviced using procedures described in paragraph 249. To service the 4.8:1 reduction ratio, heavy duty planetary reduction unit used on 399 models, refer to the following paragraph 250.

250. OVERHAUL. To service the heavy duty planetary reduction unit, first refer to paragraph 247 and remove the planetary unit, then lift off ring gear (20—Fig. 260). Drive out pinion shaft retaining pins (12), then pull pinion shafts (13) from planetary carrier (11). Carefully withdraw three pinion gears (17) with thrust washers (14), loose needle rollers (15) and spacer washers (16). Each planetary pinion gear contains two rows of loose needle rollers (15) separated by a washer (16). Each path of rollers (15) contains twenty-two individual rollers.

> CAUTION: Do not lose, damage or mix rollers from other gears and make sure to account for all rollers, making sure that none stick to gear teeth or other surface.

Slide sun gear (18) out through one of the openings in planetary carrier (11). Use special tool (MF 265A or equivalent) and a 50 ton press to remove the planetary carrier from the wheel axle shaft splines. Remove snap ring (10) from groove in axle shaft, then use a suitable press to remove wheel axle shaft (1) from the drive cover (6) and bearing cone (9).

Clean, inspect and renew any components that are damaged or worn excessively. Petroleum jelly should be used to lubricate and retain components while assembling.

Dust shield (2) is retained to wheel axle (1) by wheel studs and should not be removed unless renewal is necessary. Apply a light coat of "Hylomar" or similar sealer to outer surface of wheel axle seal (4), then install seal in drive cover (6) with metal side out. Seal should be installed flush with cover bore. Coat seal lip and fill cavity between seal and bearing cup (8) with petroleum jelly. Assemble wheel axle (1), drive cover (6) and bearing (8 and 9), pressing bearing cone (9) on shaft only far enough to permit installation of snap ring (10). After snap ring is installed, press end of axle shaft until bearing is firmly seated against

snap ring. Make sure snap ring is firmly seated in shaft groove, then press planetary carrier (11) on wheel axle shaft until it bottoms. Preassemble planet gears (17), bearing rollers (15), spacer washer (16) and thrust washers (14), using petroleum jelly. Twenty-two bearing rollers are in each path and each planet gear has two paths of bearings separated by a washer (16). Install sun gear (18) and the preassembled pinion gears in carrier (11). Insert pinion shafts (13) in carrier, aligning shaft retaining pin holes with corresponding holes in carrier, then install pins (12).

Install wheel axle and planetary assembly and adjust planetary carrier bearing preload as outlined in paragraphs 247 and 248.

DRIVE AXLE

340 Models With Direct Drive Axle

251. REMOVE AND REINSTALL. The axle shafts can be removed without removing axle housings if desired. Raise and support rear of tractor, then remove rear wheel and tire. Refer to paragraph 243 for early models with drum brake located at outer end of axle or to paragraph 245 for later 340 models with disc brake located at inner end of axle, inside axle housing.

All Models With Planetary Final Drive

252. REMOVE AND REINSTALL. To remove the drive axle (22—Fig. 259 or Fig. 260), first remove the planetary assembly as outlined in paragraph 247. Apply the handbrake fully and engage the differential lock (if right axle is being removed), then withdraw the drive axle (22) from the differential side gear, differential lock coupling and brake discs. It may be necessary to remove the complete axle housing, as outlined in paragraph 246, to reinstall drive axle if parts do not remain in alignment while drive axle is removed. Bearing cup (24) and shims (25) can be removed using a suitable puller.

BRAKES

All 340 tractors are equipped with direct drive rear axles, but early models are equipped with expanding shoe drum brakes. Brake drums are attached to the outer ends of the axle shafts on early 340 models. All tractors except 340 models are equipped with a planetary reduction unit at the outer ends of axles, and disc brake units are located at axle inner ends inside the axle housings. Late 340 tractors are equipped with

disc type brake units similar to all other tractor models.

Brakes on all models are operated hydraulically, but three types of slave cylinders and linkage are used. The slave cylinder for early 340 models with drum brakes, shown in Fig. 262, twists the brake camshaft to spread the brake shoes. On tractors with disc brakes up to serial number V39466, the slave cylinder is shown in Fig. 267. The slave cylinder used on tractors after serial number V39465, is shown in Fig. 268.

ADJUSTMENT

340 Models With Drum Brakes

253. Raise and block rear of tractor so rear wheels are off the ground. Move adjusting cover to the side and turn adjuster star wheel (3—Fig. 261) until shoes contact brake drum. Back off star wheel until drum is just free to turn, then adjust brake on other wheel the same. Latch the two wheel-brake pedals together and check to be sure that tractor does not pull to one side when pedals are depressed.

Adjust parking brake cables by turning nuts (N—Fig. 262) located on each side. Adjust cable until hand lever locks at about the third notch. Be sure that cables on both sides are adjusted evenly.

Models With Disc Brakes
Before Serial Number V39466

254. Raise and block rear of tractor so rear wheels are off the ground. Tighten adjusting nut (A—Fig. 263) until brake is locked, then loosen adjusting nut until wheel turns with only slightly perceptible drag. The nuts (A) should be backed off the same number of flats for brake assemblies on both sides. Normally the adjusting nut of 350, 355 and 360 models with three brake friction discs should be backed off five flats. The adjusting nut on 340 models with six brake friction discs and other tractor models with four friction discs should be backed off six flats. Check to make sure that brakes are adjusted equally on both sides.

Adjust parking brake cables by turning nuts (N) located on each side. Adjust cable until hand lever locks at about the third notch. Be sure that cables on both sides are adjusted evenly.

Fig. 262—Parking brake adjusting nuts (N) are located on both sides of early 340 models with drum brakes.

Fig. 261—View showing method of turning star wheel of brake adjuster on 340 models with drum brakes. Drum is removed in illustration for clarity, but drum should be installed when adjusting.

Fig. 263—View of brake actuator typical of type used on all models with disc brakes before serial number V39466.

Models With Disc Brakes
After Serial Number V39465

255. Raise and block rear of tractor so rear wheels are off the ground. Tighten adjusting nut (A—Fig. 264) until brake is locked, then loosen adjusting nut until wheel turns with only slightly perceptible drag. The nuts (A) should be backed off the same number of flats for brake assemblies on both sides. Normally the adjusting nut of 350, 355 and 360 models with three brake friction discs should be backed off five flats. The adjusting nut on 340 models with six brake friction discs and other tractor models with four friction discs should be backed off six flats. Check to make sure that brakes are adjusted equally on both sides.

Adjust parking brake cables by turning nuts (N) located on each side. Adjust cable until hand lever locks at about the third notch. Be sure that cables on both sides are adjusted evenly.

FLUID AND BLEEDING

All Models

256. The brake fluid reservoir should remain filled with a mineral (petroleum) based oil such as Massey-Ferguson Powerpart brake fluid number 3405 389 M1. Fluid level should be maintained at level marked on reservoir and not allowed to run dry.

Whenever the brake system has been drained, refill with approved fluid to the proper level, then bleed air from the system. Bleeding can be accomplished in a

Fig. 264—View of brake actuator typical of type used on all models after serial number V39465.

conventional manner using a commercially available pressure or vacuum bleeding canister. Follow manufacturer's instructions if using a commercial bleeding system. Air can also be bled from the system manually.

If equipped with trailer brake system, bleed left side at trailer brake valve first. On late models (1991 and later) with high speed gearbox (identified by "S" prefix to serial number) both left and right brakes must be bled at the same time, because a tube connects the two master cylinders to balance the braking action.

Connect a clear plastic tube to air vent screw, bleeder valve (B—Fig. 262, Fig. 263 and Fig. 264), and insert free end of tube in container with enough brake fluid to cover end of tube. Open bleeder valve and operate the brake pedal quickly, several times until reservoir is nearly empty then refill reservoir. Repeat cycle of operating the pedal and refilling reservoir until fluid expelled from tube is free of air bubbles, then close bleeder valve. It is usually necessary to cycle approximately ½ liter (1 pint) of brake fluid before all air is removed. Check for solid pedal feel and bleed again if sponginess is felt. Expelled fluid should not be reused if dirty, but clean fluid can be reused if allowed to sit overnight to remove air bubbles.

MASTER CYLINDERS

All Models

257. Either master cylinder can be removed from mounting bracket after disconnecting the operating rod clevis from pedal and hydraulic lines at connections on master cylinder. Refer to Fig. 265 and remove clevis, lock nut (1) and boot (2). Remove snap ring (4), washer (5) and operating rod (3). It may be necessary to use compressed air to remove piston assembly (6 through 13). Disassembly of piston assembly is conventional, but use extreme care to prevent damage to piston (7) and seal (6). Use only clean, approved brake fluid to clean parts.

When assembling, lubricate all parts liberally. Install washer (12) with flare away from shoulder of valve rod (11). Refer to paragraph 256 for bleeding and to paragraph 253, 254 or 255 for adjustment.

SLAVE CYLINDERS

340 Models With Drum Brakes

258. Refer to Fig. 262 for illustration. Slave cylinder can be removed after detaching operating hydraulic line, loosening handbrake cable adjustment and removing the two mounting screws. Disassembly is conventional and obvious after reference to Fig.

Fig. 265—Exploded view of master cylinder typical of all models.

1. Lock nut
2. Boot
3. Actuating rod
4. Snap ring
5. Washer
6. Seal cup
7. Piston
8. Spring retainer
9. Spring
10. Retainer
11. Valve rod
12. Washer
13. Seal
14. Master cylinder body

266. Use only clean, approved brake fluid to clean parts.

When assembling, lubricate all parts liberally. Refer to paragraph 256 for bleeding and to paragraph 253 for adjustment.

Models With Disc Brakes
Before Serial Number V39466

259. Refer to Fig. 263 for typical installation. To remove slave cylinder from left side, first remove pipes, fittings, hydraulic manifold and other interfering equipment as necessary. Remove the differential lock mechanism if removing right side unit. To remove unit from either side, remove adjusting nut (A—Fig. 263 or 1—Fig. 267), detach hydraulic brake line and cover openings to prevent the entrance of dirt. Disconnect the handbrake cable and remove the mounting screws. Disassembly is obvious after removing snap ring (7—Fig. 267), snap ring (15) and pivot pin (16). Use only clean, approved brake fluid to clean parts.

When assembling, lubricate all parts liberally. Refer to paragraph 256 for bleeding and to paragraph 254 for adjustment. Refer to paragraph 238 for adjustment of differential lock.

Models With Disc Brakes
After Serial Number V39465

260. Refer to Fig. 264 for typical illustration. Slave cylinder and actuator can be removed as follows: To remove unit from left side, first remove pipes, fittings, hydraulic manifold and other interfering equipment as necessary. Remove the differential lock mechanism if removing right side unit. To remove unit from either side, detach hydraulic brake line,

then screw plug into opening to seal opening and stop cylinder pistons from expanding and falling out while removing. A second bleeder plug can be used. Disconnect the handbrake cable, remove adjuster nut (1—Fig. 268), remove the three screws attaching actuator to axle housing, then withdraw unit from axle. Be careful not to damage boot or threads of rod (18).

Fig. 266—Exploded view of slave cylinder typical of early 340 models with drum brake.

1. Pin
2. Boot
3. Snap ring
4. Piston
5. Seal cup
6. Spring
7. Slave cylinder

Fig. 267—Exploded view of slave cylinder typical of models with disc brakes before serial number V39466. The number of friction discs (26) and separator plates (27) may be different than shown. Cross section of actuator and slave cylinder is shown in inset.

1. Adjusting nut	8. Boot	15. Snap ring	22. Return springs (4 used)
2. Washer	9. Handbrake lever	16. Pivot pin	23. Cam balls (6 used)
3. Spacer	10. Return spring	17. Housing	24. Expander plate (2 used)
4. Boot	11. Piston	18. Operating rod	25. Brake stop rod
5. Spring ring	12. Seal cup (same as 14)	19. Link pin	26. Friction discs
6. Washer	13. Spacer	20. Links	27. Steel separator plate
7. Snap ring	14. Seal cup (same as 14)	21. Anchor bolts	

Spring (13) will probably push at least one of the pistons (11) from bore when plug covering opening for the hydraulic line is removed. If necessary, air can be used to blow piston from bore. Make sure bleeder valve (B) is open, clean and operates freely in threads before installing. Make sure that threads of adjuster rod (18) are clean. Seals (12), gasket (19), boot (4), "O" ring (14), self-locking adjuster nut (1) and any questionable parts should be renewed when unit is disassembled.

Lubricate all parts liberally while assembling. Remove seals (12) from grooves of pistons (11), being careful not to leave any marks on piston. Install new seals (12) in groove with cup side toward center of cylinder. Assemble pistons and seals (11 and 12) and the spring (13) in cylinder, hold the pistons together and install bleeder valves (B) in both the bleeder port and the line connection. With both ports closed, pistons should remain together compressing spring to facilitate remainder of assembly. Coat new "O" ring (14) and boot (4) with "Hylosil" or equivalent sealer

before installing. Assemble cylinder to actuator plate and tighten retaining screws to 23-29 N·m (17-21 ft.-lbs.) torque. If removed, reattach adjuster rod (18) to bellcrank (20) and make sure that spring (21) is operating properly.

When reinstalling, make sure that sealing surfaces of axle housing and actuator plate are clean and all of gasket (19) is removed. Coat gasket surface of axle housing with "Loctite 518" or equivalent and slide protector sleeve (MF 467 or equivalent) over threads of adjuster rod (18). Slide hand brake actuator and brake slave cylinder into position, being careful not to damage boot (4) and carefully guiding cylinder and pistons (11) into place between lugs of expander plates (24). Tighten the three retaining screws to 12-16 N·m (9-12 ft.-lbs.) torque. Install adjuster nut (1) and bearing (2), then reattach hand brake cable. Remove plug and reattach hydraulic line, then bleed air from system as outlined in paragraph 256. Refer to paragraph 255 for adjustment. Refer to paragraph 238 for adjustment of differential lock.

Fig. 268—Exploded view of actuator, slave cylinder and brake discs typical of models after serial number V39465. The number of friction discs (26) and separator plates (27) may be different than shown. Cross section of slave cylinder is shown in inset.

1. Adjusting nut
2. Bearing
3. Spacer tube
4. Boot
7. Snap ring
9. Handbrake lever

10. Return spring
11. Piston
12. Seal cup
13. Spring
14. "O" ring
15. Snap ring

16. Pivot pin
17. Slave cylinder housing
18. Operating rod
19. Gasket
20. Bellcrank
21. Bellcrank return spring

22. Return springs (4 used)
23. Cam balls (6 used)
24. Expander plate (2 used)
25. Brake stop rod
26. Friction discs
27. Steel separator plate

DRUM BRAKE ASSEMBLY

340 Models With Drum Brakes

261. The brake drum (1—Fig. 269) can be removed after removing the rear wheel and the two drum retaining screws. Remove shoe retaining pins and springs (8), anchor springs (2) and anchor (4). Lift shoes (6) from backing plate with adjuster (3) and spring (7). To remove brake camshaft, loosen set screw in brake operating lever, then withdraw camshaft from backing plate (10) and remove lever from inner end.

When reassembling, lubricate all parts liberally and reverse disassembly procedure. Set screw in camshaft operating lever should be tightened to 1.4-3.4 N•m (12-30 in.-lbs.) torque, anchor pin retaining nuts should be tightened to 200 N•m (150 ft.-lbs.) torque. Refer to paragraph 253 for adjustment.

Fig. 269—Exploded view of shoe and drum type brakes used on early 340 models. Later 340 models and other models are equipped with disc brakes.

1. Brake drum
2. Anchor springs
3. Adjuster
4. Anchor
5. Anchor pins

6. Brake shoes
7. Adjuster spring
8. Hold down unit
9. Brake camshaft
10. Backing plate

DISC BRAKE ASSEMBLY

All Models With Disc Brakes

262. Refer to paragraph 244 for removing direct drive rear axle housing from late 340 models with disc brakes or paragraph 246 for removing rear axle housing from models with planetary reduction. Refer to paragraph 259 or paragraph 260 and remove the brake slave cylinder and actuator unit shown in Fig. 267 or Fig. 268. Refer to paragraph 239 for removal of differential lock assembly. Refer to Fig. 255 and remove screws (S) that attach the differential carrier housing (1) to the axle housing. Remove carrier housing and refer to Fig. 267 or Fig. 268.

On early models, remove pin (19—Fig. 267) and operating rod (18). Remove brake stop rod (25) and discs (26), plates (27) and expander assembly (20, 21, 22, 23 and 24). Models 350, 355 and 360 tractors are equipped with three friction discs (26) and one steel plate (27) assembled as shown. Other models are equipped with four friction discs (26) and two steel plates equally divided between sides of expander assembly. Depth of groove in friction disc should be more than 0.3 mm (0.012 inch).

On later models, detach rod (18—Fig. 268) from lever (20). Remove brake stop rod (25) and discs (26), plates (27) and expander assembly (20, 21, 22, 23 and 24). Model 340 tractors are equipped with six friction discs (26) and three steel plates (27). One steel plate and two friction discs are located between wheel end of axle and expander assembly; four friction discs and two steel plates are located between expander and differential carrier housing. Models 350, 355 and 360 tractors are equipped with three friction discs (26) and one steel plate. Two friction discs and the one steel plate are located between wheel and expander. Other models are equipped with four friction discs (26) and two steel plates equally divided between sides of expander assembly. Depth of groove in friction disc should be more than 0.3 mm (0.012 inch). Depth of groove in friction disc should be more than 0.3 mm (0.012 inch).

Reassembly is reverse of disassembly for all models. Make sure that friction discs and steel plates are installed in order described. Refer to paragraph 239 for installation of differential lock assembly and to paragraph 259 or paragraph 260 for installation of the brake slave cylinder and actuator unit. Refer to paragraph 244 for installing direct drive rear axle housing on late 340 models or paragraph 246 for installing axle housing on models with planetary reduction. Refer to paragraph 256 and bleed brakes, then refer to paragraph 254 or 255 to adjust brakes.

LIVE POWER TAKE-OFF

263. All models are equipped with either a dual or split torque clutch. The dual clutch controls engagement of drive to the engine transmission and to the live power take-off (pto). Split torque clutches are installed on tractors equipped with independent pto. Refer to the appropriate following paragraphs for service to live pto.

OUTPUT SHAFT, SEAL AND REAR BEARING

Models With Live Pto

264. To remove pto output shaft, drain oil from axle center housing and transmission, shift pto control to engaged live position and remove pto shield. Detach bottom of 3-point hitch center control beam, remove retainer attaching nuts (2 and 4—Fig. 270), then remove the two center screws. Unbolt and remove retainer (3) and plate (5), then use threaded cap (1) to pull seal retainer (7) from center housing. Shaft (9) will probably be removed with seal and rear bearing.

To install pto output shaft assembly, press bearing (10) onto shaft and seal into housing (7). Rear of seal (6) should be 1 mm (0.04 inch) below edge of housing. Renew "O" ring (8), coat seal (6) and "O" ring liberally with transmission oil and install shaft, bearing and seal into center housing. Drive seal housing into center housing and install retainer (3).

Tighten the four nuts with spacers (4) to 113-169 N·m (83-125 ft.-lbs.) torque and the two nuts (2) attaching drawbar to chain bracket to 230-260 N·m (170-192 ft.-lbs.) torque.

Output shaft front needle bearings can be renewed after first removing the differential assembly as outlined in paragraph 235. Bearing should be removed toward the rear. Use an appropriate size driver to

Fig. 270—View of pto output shaft
and associated parts for models
with live pto.
1. Cap
2. Nuts to drawbar bracket
3. Retainer
4. Nuts, washers and spacers
5. Seal housing plate
6. Triple lip seal
7. Seal housing
8. "O" ring
9. Output shaft
10. Rear bearing
11. Snap ring

install new needle bearing just below flush with bore at rear.

ENGAGEMENT COUPLING AND GROUND SPEED GEARS

Models With Live Pto

265. To remove engagement coupling, two speed gears or ground speed gears, remove rear shaft as outlined in paragraph 264, then refer to paragraph 240 for access to the coupling and, if so equipped, the ground speed drive gears.

PTO FRONT DRIVE SHAFTS AND GEARS

Models With Live Pto

266. Refer to the appropriate transmission section for removing or servicing pto drive shafts, gears or bearings which transfer drive from the engine clutch to the pto engagement coupling.

INDEPENDENT POWER TAKE-OFF

267. All models are equipped with either a dual or split torque clutch. Split torque clutches are installed on tractors equipped with independent pto. Drive for the hydraulic pump and for the pto are taken from the splined hub of a plate attached to the clutch cover, which turns all the time that engine is running. Control of pto engagement or disengagement is by a separate, hydraulically operated clutch. The dual clutch controls engagement of drive to the engine transmission and to the live power take-off (pto). Refer to the appropriate following paragraphs for service to independent pto.

OPERATION

Models With Independent Pto

268. Some tractors may be equipped with an independent pto which is driven by a flywheel "Split Torque" clutch and controlled by a hydraulically actuated multiple disc clutch contained in the differential housing. The pto front drive shaft is splined into hub of clutch cover and turns continuously when engine is running.

The pto control lever is on the left side cover of differential housing. Moving lever forward disengages the hydraulically actuated multiple disc clutch and engages the pto brake. Moving lever rearward releases the hydraulic brake and engages the multiple disc pto clutch.

Hydraulic pressure to actuate the pto clutch and brake is produced by the auxiliary gear type pump which also provides power for the auxiliary hydraulic system if so equipped.

OUTPUT SHAFT

Models With Independent Pto

269. REMOVE AND REINSTALL. To remove the output shaft (9—Fig. 271), first drain oil from axle center housing and transmission and remove pto shield. Detach bottom of 3-point hitch center control beam, remove retainer attaching nuts (2 and 4), then remove the two center screws. Unbolt and remove retainer (3) and plate (5), then use threaded cap (1) to pull seal retainer (7) from center housing. Shaft (9) will probably be removed with seal and rear bearing.

Fig. 271—View of pto output shaft and associated parts for models with independent pto.

1. Cap
2. Nuts to drawbar bracket
3. Retainer
4. Nuts, washers and spacers
5. Seal housing plate
6. Triple lip seal
7. Seal housing
8. "O" ring
9. Output shaft
10. Rear bearing
11. Snap ring
12. Bearing sleeve
13. "O" ring
14. Snap ring

To install pto output shaft assembly, press bearing (10) onto shaft and seal into housing (7). Rear of seal (6) should be 1 mm (0.04 inch) below edge of housing. Renew "O" ring (8), coat seal (6) and "O" ring liberally with transmission oil and install shaft, bearing and seal into center housing. Drive seal housing into center housing and install retainer (3).

Tighten the four nuts (4) with spacers to 113-169 N·m (83-125 ft.-lbs.) torque and the two nuts (2) attaching drawbar to chain bracket to 230-260 N·m (170-192 ft.-lbs.) torque.

Output shaft front needle bearings can be renewed after first removing the differential assembly as outlined in paragraph 235. Bearing should be removed toward the rear. Use an appropriate size driver to install new needle bearing just below flush with bore at rear.

LEFT SIDE COVER AND CONTROL VALVE

Models With Independent Pto

270. REMOVE AND REINSTALL. Drain transmission until level of oil is at low mark on dipstick. Detach control rods and hydraulic lines that would interfere with removal of side cover and remove or relocate as necessary. Remove plug (1—Fig. 272) and locator plug (2) and nuts (3 and 4). Unbolt and remove cover, making sure that tubes do not catch.

When installing, install new gasket, "O" rings and sealing washers. Be sure that internal lever engages the control valve. Coat gasket face of side cover with "Loctite 515 Instant Gasket" or equivalent. Reassemble by reversing removal procedure. Seal the six screws retaining side cover with "Hylomar" or equivalent sealer.

PTO CLUTCH, BRAKE AND CONTROL VALVE

Models With Independent Pto

271. REMOVE AND REINSTALL. To remove the clutch and control valve unit, first remove hydraulic lift cover as outlined in paragraph 285 and split

Fig. 272—Exploded view of left side cover showing hydraulic lines and valve control lever.

1. Plug
2. Locator plug
3. "O" ring
4. "O" ring
5. Nut
6. Nut
7. "O" ring
8. Washer
9. Tube
10. Tube
11. Gasket
12. Snap ring
13. "O" ring
14. Washer
15. Control valve operating lever
16. Left side cover
17. Plate

tractor between differential housing and transmission housing as outlined in the appropriate paragraph in the transmission section. Remove pto left side cover, disconnect hydraulic lines and remove pump support pins, then remove hydraulic pump and independent pto clutch unit through front of differential housing.

To reinstall clutch assembly, reverse the removal procedure.

272. OVERHAUL. Refer to Fig. 273 for an exploded view of clutch, brake and control valve assembly. To disassemble the removed clutch, first remove snap ring (20), thrust washer (19) and control valve housing (18) from clutch hub end of housing (15). Remove retaining ring (4) and withdraw retainer plate (1), thrust washer (2), drive hub (3), shims (5), wave springs (6), separator plates (7) and friction plates (8) from clutch housing. Tap clutch housing against wooden bench to remove piston front plate (10), Belleville washer (11), spacer (13) and clutch piston (14).

To disassemble control valve, remove detent plug, spring and ball (21). Drive out roll pin (23), then withdraw plunger (33) and valve spool (27) assembly from valve body. Unseat snap ring (32) and separate valve spool from plunger. Drive roll pin (29) from spring guide (30) and remove guide and spring (28) from valve spool. Remove retaining ring (24), then tap sleeve (26) with "O" ring from valve body bore.

Unbolt and remove brake pad (40) and brake cylinder (37) from control valve housing. Remove retaining ring (34) and withdraw piston (39) from cylinder.

Renew all "O" rings and sealing rings. Renew clutch separator rings and friction discs if excessively worn, warped or discolored from overheating. Free height of wave springs (6) should be 5.21 mm (0.205 inch) and all seven should be within 0.51 mm (0.020 inch) of the same height. If new friction discs are installed, soak discs in transmission oil for at least 30 minutes before assembling.

When assembling pto brake, a new shoulder bolt (41) should be installed to retain brake pad. Tighten shoulder bolt and other screw to 7 N·m (60 in.-lbs.) torque. Be sure that brake pad moves freely.

To reassemble control valve, reverse the disassembly procedure. Insert roll pin (23) with split side of pin facing away from plunger (33). Tighten detent plug

Fig. 273—Exploded view of typical ipto multiple disc clutch, control valve assembly and brake assembly.

1. Retainer and gear
2. Thrust washer
3. Drive hub
4. Retaining ring
5. Shims
6. Wave springs (7 used)
7. Separator plates (7 used)
8. Friction discs (7 used)
9. Piston seal
10. Piston front plate
11. Belleville washer
12. Piston seal
13. Spacer
14. Clutch piston
15. Clutch housing
16. Check valve assembly
17. Seal rings (2 used)
18. Control valve housing
19. Thrust washer
20. Snap ring
21. Detent spring & ball
22. Ball
23. Roll pin
24. Retaining ring
25. "O" ring
26. Sleeve
27. Valve spool
28. Spring
29. Pin
30. Spring guide
31. Washer
32. Retaining ring
33. Plunger
34. Retaining ring
35. Spring seat
36. Return spring
37. Brake cylinder
38. "O" ring
39. Piston
40. Brake pad
41. Shoulder bolt

to 47 N·m (35 ft.-lbs.) torque. Piston rings (17) are installed in grooves of housing (15), but should have end gap of 0.050-0.305 mm (0.002-0.012 inch) in bore of housing (18).

Lubricate piston seal rings with clean transmission/hydraulic oil when assembling. Install piston, flat side first, and spacer ring into clutch housing. Install Belleville washer with concave (dished) side outward, then install piston front plate with stepped edge inward. Install the seven steel plates (7), seven wave springs (6) and seven friction discs (8) alternately, beginning with an externally splined steel separator plate (7), followed by a wave spring (6) and finally a friction disc (8). New friction discs should be soaked in oil before assembling. To facilitate assembly, small pins can be inserted through the three vent holes in clutch housing (15) to hold plates and springs compressed. Remove pins before completing assembly.

Select proper thickness of shims (5) as follows. Assemble retainer plate (1) over the last sintered friction disc (8), leaving out shims (5) and thrust washer (2), then install snap ring (4). Press retainer (1) into housing (15) with a force of about 64 kg. (140 lbs.) to compress wave springs, then measure distance between retainer (1) and snap ring (4). Gap

should be 2.54-3.05 mm (0.100-0.120 inch). If gap exceeds 3.05 mm (0.120 inch), install 0.20-0.51 mm (0.008-0.020 inch) or 0.30-0.76 mm (0.012-0.030 inch) thick shims as necessary. If available, an extra retainer plate (1) can be modified by machining gear away to provide more room for measuring clearance. **The modified retainer should be used only for measuring and should not be installed.** Remove snap ring (4) and retainer plate (1), then install thrust washer (2) and selected shims (5). Reinstall retainer and be sure that snap ring is fully seated in groove.

HYDRAULIC PRESSURE TESTS

Models With Independent Pto

273. Remove test plug (1—Fig. 272) from left side cover and attach a suitable 3000 kPa (400 psi) pressure test gauge to the port. Start engine and run until transmission oil reaches normal temperature. Set engine speed to 1200 rpm, engage pto and observe pressure indicated at test gauge. Pressure should be 1550-1900 kPa (225-275 psi) for models with single speed pto.

HYDRAULIC SYSTEM

274. All models are equipped with basically two separate hydraulic systems: A main hydraulic system and an auxiliary hydraulic system. Additionally, the auxiliary system is divided into two separate systems, one supplying pressure for steering, ipto clutch and brake, the second supplying pressure for external cylinders and spool valves.

The main hydraulic system consists of a four-piston-type pump submerged in hydraulic fluid in the differential housing and driven by the pto drive shaft. A control valve is located in the pump unit which meters the operating fluid at pump inlet. A single-acting hydraulic cylinder, attached to bottom of lift cover, actuates the hitch rockshaft.

The two section, gear type, auxiliary pump is installed on the right side of the engine and driven by the engine gears or auxiliary drive shaft. Oil is filtered through a 140 micron inline filter before entering either of the two pump sections. One pump section provides pressure for steering and ipto. The second pump section delivers pressurized oil to the auxiliary hydraulics.

Refer to the appropriate following paragraphs for service to main hydraulic system or to paragraphs 293 through 297 for service to the auxiliary hydraulic system.

RESERVOIR AND FILTERS

All Models

275. The transmission, transfer drive (four-wheel-drive models), differential, rear axle, pto and hydraulic systems share the same fluid. Transmission housing, center housing and rear axle housings serve as system reservoir and are each provided with individual drain plugs. Recommended fluid is "Massey-Ferguson Super 500" multi-use 10W-30 oil, "Massey-Ferguson Permatran" or equivalent oil. Capacity is approximately 43.4 L (11.5 gal.) **without** spacer or four-wheel drive or 47.4 L (12.5 gal.) **with** spacer or four-wheel drive. Specific options may add or reduce the amount of oil and, if drained, varying amounts of oil will cling to internal parts and walls of housings. Maintain oil at full level of dipstick shown in Fig. 274 or Fig. 275.

The main hydraulic system inlet strainer, located in bottom of rear axle center housing, should be removed and cleaned or renewed every 1000 hours of operation. Drain all oil from the transmission and rear axle center housings, then remove the three screws retaining cover (1—Fig. 276) to bottom of center housing. Remove clip (3), nut (4), filter (8) and

Fig. 275—Transmission oil level of tractors with cab is measured by dipstick that points toward rear as shown.

Fig. 274—Transmission oil level of tractors without cab is measured by dipstick located in the side cover as shown. Refer to Fig. 275 for models with cab.

directed to the auxiliary manifold. This canister type filter should be removed and new filter installed after each 250 hours of operation. Wipe seal at top of canister with oil before installing, then run engine and check for leaks.

TROUBLESHOOTING

All Models

276. The following is a list of symptoms which may occur during operation and their possible causes. These suggestions can be used in conjunction with other Tests and Adjustments to isolate the cause of problems before proceeding with component disassembly.

1. Hitch will not raise. Could be caused by:
    a. Internal oil leak in system. Remove response control side cover and check for leakage.
    b. Damaged, binding or maladjusted control valve linkage.
    c. Maladjusted or faulty servo valve (pressure control units) or faulty safety relief valve (non-pressure control units).

remaining parts from opening. Clean all parts thoroughly and install new gasket and "O" rings. Reinstall inlet filter assembly and drain plugs, then refill with clean approved oil.

Oil entering the two section auxiliary pump is screened by filter (4—Fig. 277) located in the suction line. This filter should be removed, cleaned or renewed and reinstalled every 1000 hours of operation. Drain approximately 10 L (2 gal.) of oil from transmission housing, before unbolting and removing cover (1). Install new gaskets, then tighten retaining bolts evenly and securely.

Pressurized oil not required for operating the power steering system or ipto clutch is circulated through a cooler and filter (F—Fig. 278) before being

Fig. 276—Main hydraulic system inlet strainer should be removed and cleaned or renewed every 1000 hours of operation. A large "O" ring, not shown, is located above tube (9).

1. Cover
2. Gasket
3. Lock clip
4. Nut
5. Spring
6. Washer
7. "O" ring
8. Filter
9. Tube

Fig. 277—Inlet screen shown is located on inlet line to auxiliary pump and filters oil entering both steering and auxiliary sections of pump.

1. Cover
2. Spring
3. Gasket
4. Filter

d. Faulty main hydraulic pump.
2. Lift links do not raise evenly. Jerky operation could be caused by:
   a. Valve sticking in pump valve chamber.
3. Lift links will not raise to full height. Could be causes by:
   a. Transport stop maladjusted.
   b. Control valve maladjusted.
   c. Control linkage maladjusted.

Fig. 278—Canister filter (F) should be renewed every 250 hours of operation.

4. Lift links will not lower. Could be caused by:
   a. Control valve sticking.
   b. Control valve maladjusted.
   c. Lift arms binding.
5. Hitch lowers too fast with response control in "Slow" position. Could be caused by:
   a. Response control maladjusted.
   b. Control linkage binding or damaged.
6. Erratic action when operating in draft control. Could be caused by:
   a. End play in master control spring.
   b. Control linkage binding or damaged.
7. Lift links creep down. Could be caused by:
   a. Internal leakage in lift cylinder, control valve or valve chambers.

TESTS AND ADJUSTMENTS

All Models

277. All operational tests should be conducted with hydraulic fluid at normal operating temperature of about 50-70° C (120-160° F). Always cover open lines and ports as soon as possible to prevent the entrance of dirt and unnecessary fluid loss.

CAUTION: To prevent injury, engine should be stopped, all hydraulically actuated equipment should be rested on the ground and all hydraulic pressure within circuits relieved before removing any test port or disconnecting any hydraulic line.

278. PRESSURE. Remove test plug from port (P—Fig. 279) on left side of lift cover and connect a suitable 30,000 kPa (4000 psi) pressure test gauge to the port. If equipped with selector valve, move control lever to "Linkage" position. Start engine and operate at 1200 rpm, move the Position Control lever to "Constant Pumping" position and observe pressure indicated by test gauge. Correct pressure is 22,100-

Fig. 279—Exploded view of master control spring assembly used on all models. Main pump pressure may be tested at port (P).

F. Filler opening
P. Test port
1. Head
2. Pin
3. Boot
4. Adjusting nut
5. Control spring
6. Spring guide
7. Plunger
8. Overload stop disc
9. Set screw
10. Nylon plug
11. Lift cover
12. Breather
13. Standpipe
14. "O" ring
15. Washer

23,100 kPa (3200-3350 psi) for all models. Some models (from serial number N42314, manufactured since November, 1988) are equipped with a non-adjustable cartridge. Relief valve setting is adjustable on other models. Relief valve is accessible through opening after removing cover from right side of rear axle center housing.

279. MASTER CONTROL SPRING. Move the draft control lever to "Down" position. Disconnect control beam from spring head (1—Fig. 279) and pivot beam down to expose spring head. Check for end play by pulling and pushing spring head. If end play is indicated, remove Allen set screw (9) from side of housing and pull boot (3) back to expose adjusting nut (4). Unscrew nut (4) from lift cover using special spanner wrench and withdraw spring assembly from top cover. Nylon plug (10) will be damaged and should be removed so that new plug can be installed.

Hold spring head (1) and attempt to turn spring (5). Spring should fit snugly with no end play, but should still turn with moderate effort. If adjustment is incorrect, drive pin (2) out and thread spring head (1) onto or from plunger (7) until end play is just eliminated. Tighten plunger as required until slot is aligned and reinstall pin (2). Reinstall control spring assembly in lift housing bore and tighten nut (4) until end play is just eliminated. If adjusting nut is either too loose or too tight, unit will have noticeable end play. Install new nylon plug (10) and tighten Allen set screw (9) to 7 N•m (60 in.-lbs.) torque when adjustment is correct.

280. EXTERNAL LINKAGE ADJUSTMENTS. The following adjustments should be performed with the lift cover installed and the tractor operational. Loosen quadrant bolts (1—Fig. 280), locate quadrant in center of slotted hole, then tighten bolts. Move the Draft Control lever (2) to between "Sector" marks and Position Control lever (3) to the "Transport" position. Locating holes (5) for position control and locating holes (6) for draft control should both be aligned so that 6 mm (¼ inch) diameter pins can be inserted. Linkage attaching holes (N) are used if tractor is not equipped with spacer or four-wheel-drive transfer case. Linkage is attached to holes (S) if spacer is located between transmission and rear axle center housings. Remove pins from holes (5 and 6) after adjusting lengths of control rods (7).

Attach a weight of at least 400 kg (900 lbs.) to the lower lift links, start engine and set speed to 1200 rpm. Top link should not be attached. Cycle controls throughout operating range at least six times to bleed air from system. Move the Position Control lever (3) to "Transport" and the Draft Control lever (2) between "Sector" marks. Loosen quadrant mounting bolts (1) and move the draft control quadrant and lever until the lower links are held in horizontal position. Tighten quadrant mounting bolts (1) and recheck alignment of holes (5 and 6) as described in previous paragraph.

Move Draft Control lever to the fully UP position and the Position Control lever to "Transport." Loosen quadrant mounting bolts (1) and rotate the position control quadrant and lever until the rear linkage reaches maximum travel (indicated by relief valve

Fig. 280—Views of Position Control lever, Draft Control lever, related linkage and adjustment points. Refer to text for adjustment procedure.

1. Quadrant mounting bolts
2. Draft Control lever
3. Position Control lever
4. Adjustable stop
5. Locating hole for Position Control
6. Locating hole for Draft Control
7. Control rods

Fig. 281—Marks should be scribed on lift arms and lift cover as described in text to measure movement of lift arms when adjusting linkage. Distance (D) between marks on cover should be 6 mm (1/4 inch).

Fig. 282—Refer to text for adjusting response lever (R).

opening). Scribe a mark (A—Fig. 281) on lift cover and a similar line (B) on lift arm. Scribe a second mark (C) on lift cover 6 mm (¼ inch) to the rear and above the first mark. Rotate the position control quadrant and lever until the lift arms lower slightly and mark (B) on lift arm aligns with mark (C) on cover. Tighten bolts (1—Fig. 280) and recheck alignment of holes (5 and 6) as described in earlier paragraph.

Move the response lever (R—Fig. 282) so that it is 6 mm (¼ inch) from slow position as shown at (D). Loosen screw (S), rotate inner cam clockwise by mov-

ing the locking screw (S) in slot until it just contacts the internal response lever on hydraulic pump, then tighten screw (S). Connect control rod and check "Response" Control operation.

281. INTERNAL LINKAGE ADJUSTMENTS. The following adjustments to internal lift linkage must be accomplished before installing lift cover. Special tools are required.

Invert the removed lift cover in a fixture or on bench, blocking cover so that rockshaft can be moved to normal and raised transport positions. Move draft control link (D—Fig. 283) to fully raised position and

Fig. 283—Clearance between adjusting screw (43) and lift cover should be 5.8 mm (0.230 inch) and can be measured using the correct size gauge as shown at (G).

Fig. 284—View showing adjustment of draft control neutral setting for tractors without intermix (All except some 399 tractors). Special tool MF 273 (8) is used to attach load (9) to the end of vertical lever (10). Special tool MF 356C (11) sets the vertical lever to the correct (pump neutral) position.

make sure that draft control rod is against spring plunger by the linkage return spring. Clearance between adjusting screw (43) and lift cover should be 5.8 mm (0.230 inch) and can be measured using the correct size gauge as shown at (G).

282. To adjust draft control neutral setting for tractors **without intermix** (All except some 399 tractors), proceed as follows: Set the draft control link to "Sector" position and insert a 6 mm (1/4 inch) diameter pin (6—Fig. 284). Move position control link to "Transport," then insert a 6 mm (1/4 inch) diameter pin (5). Loosen socket head screw (7) and locknut on vertical lever to ensure that linkage does not interfere. Attach special tool (MF 273 or equivalent) to lift cover and attach a 1.3 kg (3 lbs.) load (9) to the end of vertical lever (10). The load (9) simulates the force of the linkage pump control spring. Attach special tool (MF 356C or equivalent) as shown at (11) to set the vertical lever to the correct (pump neutral) position. Loosen nut (29) and adjust position of lever pivot until lever just contacts the horizontal pin of special tool (11), then tighten nut (29). Recheck adjustment, then bend lock tab against nut (29). Refer to paragraph 284 for setting position control neutral ("Transport") setting.

283. To adjust draft control neutral setting for tractors **with intermix** (Some 399 tractors), proceed as follows: Set the draft control link to "Sector" position and insert a 6 mm (1/4 inch) diameter pin (6—Fig. 285). Move position control link to "Transport", then insert a 6 mm (1/4 inch) diameter pin (5). Loosen socket head screw (7) and locknut on vertical lever to ensure that linkage does not interfere. Attach special tool (MF 445) as shown at (13) to correctly position the lift arms. Marks installed at the factory on the lift arm and lift cover should be aligned when the special tool is installed. These marks on lift arm and cover can be used to position the lift arms without using the special tool if carefully done, and a retaining plate screw can be used to hold the arms in position.

Attach special tool (MF 273 or equivalent) to lift cover as shown at (8) and attach a 1.3 kg (3 lbs.) load (9) to the end of vertical lever (10). The load (9) simulates the force of the linkage pump control spring and a straight spring scale pull can be attached to lever (10) to apply the correct force.

Attach special tool (MF 356C or equivalent) as shown at (11) to set the vertical lever to the correct (pump neutral) position. Loosen nut (29) and adjust position of lever pivot until lever just contacts the

Fig. 285—View showing adjustment of draft control neutral setting for tractors with intermix (Some 399 tractors). Special tool MF 273 (8) is used to attach load (9) to the end of vertical lever (10). Special tool MF445 (13) can be used to set the position of the lift arms. Special tool MF 356C (11) sets the vertical lever to the correct (pump neutral) position.

Fig. 286—Draft control link (6) is fully UP in position shown. Position control link is in "Transport" location when pin (5) is installed through locating holes.

horizontal pin of special tool (11), then tighten nut (29). Recheck adjustment, then bend lock tab against nut (29). Refer to paragraph 284 for setting position control neutral ("Transport") setting.

284. The following procedure for setting position control neutral setting is used for both tractors with and without intermix. Position the draft control link in fully UP position as shown at (6—Fig. 286). Draft control link does not have alignment holes in the fully UP position. Move position control link to "Transport," then insert a 6 mm (¼ inch) diameter pin (5).

Rotate the cam ram arm against special tool (MF 272 or equivalent). If special tool is not available, marks (B and C—Fig. 281) can be aligned. Refer to paragraph 280 for correct placement of marks. Lift arm retaining plate screw can be used to hold the arms in position and simulate installation of special tool (MF 272).

Attach special tool (MF 273 or equivalent) to lift cover as shown at (8—Fig. 286) and attach a 1.3 kg (3 lbs.) load (9) to the end of vertical lever (10). The

load (9) simulates the force of the linkage pump control spring and a straight spring scale pull can be attached to lever (10) to apply the correct force.

Attach special tool (MF 356C or equivalent) as shown at (11) then set position of the vertical lever (10) by turning screw at (14) until lever just contacts the horizontal pin of special tool (11), then tighten adjustment locknut. Recheck adjustment, after tightening locknut.

LIFT COVER

All Models

285. REMOVE AND REINSTALL. Disconnect lift links from lift arms and remove upper link control beam from master control spring clevis.

On models without cab, disconnect hydraulic lines attached to the spool valve block, disconnect spool valve control rods, then remove the spool valve. Disconnect the pto control rod to the left side cover, remove the four-wheel-drive control lever (if so equipped) and disconnect the response control rod from the right side cover. If so equipped, disconnect control lever and rod from right side cover. Disconnect

Fig. 288—Attach hoist (H) as shown for removing lift cover from models with cab.

draft and position control rods (7—Fig. 280) from control levers and detach wiring harness from under the seat. Unbolt seat panel from fenders and front vertical panel, loosen screws attaching fenders to rear axles, then lift complete seat panel and control levers from tractor.

On models with cab, disconnect spool valve control rods at front end, then remove the spool valve. Unbolt and remove the complete seat panel. Remove the selector control valve lever and disconnect control rod from right side cover, if so equipped.

On all models, remove transfer cap and stand pipe or selector valve and stand pipe. Move draft control lever in full UP position as shown at (6—Fig. 286) and the position control lever in transport position and lock both levers in this position. Unbolt lift cover from housing and attach hoist to cap or sector valve as shown in Fig. 288.

Refer to proceeding paragraphs 281 through 284 for internal linkage adjustments. Clean sealing surfaces of center housing and lift cover and coat mating surfaces with "Loctite 515 Instant Gasket" or equivalent sealer. Install at least two 7/16 inch UNC studs approximately 2 inches (50 mm) long in center housing before installing cover to facilitate alignment. Be especially careful when lowering cover that vertical

lever of cover does not catch on internal components and bend. Tighten retaining screws to 88 N·m (65 ft.-lbs.) torque. Refer to paragraphs 278 through 280 for external adjustments. Remainder of assembly is reverse of disassembly procedure.

286. OVERHAUL. It should be noted before disassembling the removed lift cover that special tools may be required to adjust linkage when reassembling. Refer to paragraphs 277 through 284 for some adjustment notes.

To disassemble, loosen Allen set screw (9—Fig. 279) and unscrew adjusting nut (4), then withdraw parts (1 through 8) from housing (11). Release lock plate from around nut (29—Fig. 289), then remove nut, tab washer and spacer. Release spring (41), detach pivot bracket (31) from screw, then remove spring (38). Guide rods for control cam return springs (34) are drilled near loose end to assist disassembly and reassembly. Compress springs (34) and insert pins through rods to hold springs (34) compressed, then unbolt and remove ram cylinder (17) with bracket (30) attached. Always install new "O" ring (18) when assembling. Thread a 10-32 UNF screw into support retaining pin (7—Fig. 290), then pull or pry pin from lift cover. Hold control fingers (32 and 37—Fig. 289) away from cams (39 and 40) as support (3—Fig. 290) and shafts (1 and 2) are withdrawn. Cams (39 and 40—Fig. 289) can be removed after removing set screw and pivot pin. Rocker arms and shaft can be removed and new bushings installed if required.

Assemble lift cover using all new "O" rings and tab (locking) plates. Concave side of Belleville washer should be toward lift cover. When installing connecting rod (21) in rockshaft arm (26), coat threads of set screw (33) with "Loctite 545" or equivalent, tighten set screw until bottomed against connecting rod, then back screw up 1/4 turn. Tighten the two front nuts retaining cylinder (17) to 200-240 N·m (148-177 ft.-lbs.) torque and the two rear nuts to 280-330 N·m (207-243 ft.-lbs.) torque. Refer to paragraph 285 for installation and to paragraphs 277 through 284 for adjustment notes.

SELECTOR VALVE

Models So Equipped

287. Some models are equipped with a selector control valve (Fig. 292) which allows operation of lift linkage and auxiliary hydraulics simultaneously, or allows only the operation of auxiliary hydraulics. With the selector valve forward, in "LINKAGE" position, the lift linkage and auxiliary hydraulics may be operated independently. With selector lever back, in "EXTERNAL" (auxiliary) position, the flow of oil from the hydraulic lift pump and auxiliary hydraulic pump is combined to provide maximum flow at the remote

Fig. 289—Exploded view of typical hydraulic lift linkage, rockshaft and associated parts.

17. Cylinder
18. "O" ring
19. Piston rings
20. Piston
21. Connecting rod
22. Lock clip
23. Lift arm

24. "O" ring
25. Bushings
26. Rockshaft arm
27. Rockshaft
28. Lift arm
29. Nut
30. Support bracket

31. Pivot bracket
32. Vertical lever
33. Set screw
34. Springs
35. Cam roller
36. Position control
    adjusting screw

37. Position control finger
38. Spring
39. Position control cam
40. Draft control cam
41. Spring
42. Draft control rod
43. Stop screw

Fig. 290—Exploded view of typical control shafts, support and lift cover.

1. Draft control shaft
2. Position control shaft
3. Support
4. "O" ring
5. Position control lever
6. Draft control lever
7. Support retaining pin
11. Lift cover

couplers. The lift linkage is locked and hitch can not be used.

288. LINKAGE ADJUSTMENT. To adjust linkage, first move draft control and position control levers to "DOWN" position. Disconnect control linkage from selector valve bellcrank (7—Fig. 291). Rotate bellcrank clockwise until internal resistance is felt. Adjust length of lower control rod clevis (8) by turning clevis (8) until retaining pin (9) fits freely into

clevis and control valve bellcrank, then lengthen the rod by one turn of the clevis and install the pin.

Turn adjusting screw (6) as required so that lever just snaps into position and requires slight force. Tighten locknut when adjustment is correct.

289. R&R AND OVERHAUL. To remove the selector valve, disconnect control linkage and hydraulic pipes from valve, then remove both socket head screws retaining valve to lift cover. Lift valve from lift cover.

To disassemble, unbolt and remove end plate (30—Fig. 292). Remove dowel (D), then remove actuator shaft (26) and fork (29). Spool (28) can be removed from cover end, but plug and spring should also be removed. Remove plugs, shims (22), springs and relief valve (23). Relief valve is used to protect lift cylinder from shock loads when transporting heavy mounted implements over rough terrain with valve in external (auxiliary only) position.

Clean all parts and be sure that all old "O" rings and gaskets are removed. Clean all oil passages using compressed air.

Reassemble selector valve, reversing disassembly procedures. Relief valve (23) should open at 262-278 bar (3800-4000 psi). Increasing spring pressure by adding 0.25 mm (0.010 inch) thickness to shims (22) will increase pressure approximately 7 bar (100 psi). Relief valve opening pressure can be checked using

Fig. 291—View of right side cover (17) and linkage. Bellcrank (7) is also shown in Fig. 292.

4. Control lever	
5. Pivot	12. Bushing
6. Adjusting screw	13. Arm
7. Bellcrank	16. Response control lever
8. Clevis	17. Side cover
9. Retaining pin	18. Shaft
10. Lower control rod	19. Spring
11. Lockout lever	20. Response cam

Fig. 292—Exploded view of selector valve assembly used on some models. Bellcrank (7) is also shown in Fig. 335.

7. Bellcrank	26. Actuator shaft
22. Shims	27. Selector valve body
23. Safety relief valve	28. Valve spool
poppet and ball	29. Shift fork
24. Sleeve	30. End plate

appropriate test fixtures and pressure source above the test range.

Reinstall selector valve using new gasket and "O" ring. Tighten retaining socket head screws to 34-54 N.m (20-40 ft.-lbs.) torque. Adjust control linkage as outlined in paragraph 280.

HYDRAULIC LIFT PUMP

All Models

290. REMOVE AND REINSTALL. The lift pump can be removed from models **equipped with only single speed (540 rpm), ground speed pto**, through the top opening of differential housing after removing the lift cover. However, many mechanics prefer to split the tractor between transmission and differential housing, then remove the pump out front opening of differential housing as outlined in paragraph 291.

To remove the lift pump from tractors which can have pump removed through the top, drain transmission oil and remove lift cover as outlined in paragraph 285. Remove the pto output shaft and side covers from both the left and right sides of axle center housing. Remove split pin from center of drive shaft coupling, then remove the coupling and drive shaft. Unbolt and remove the two retaining screws or stud nuts and the pump supports (S—Fig. 293) from each side of axle center housing, then carefully work pump assembly up out of center housing.

Install by reversing removal procedure. Be sure that all old gaskets and "O" rings are removed completely and surfaces are cleaned. Use new gaskets and "O" rings when assembling. Install new split pin in drive shaft coupling hole that limits drive shaft end play to 0.38-2.54 mm (0.015-0.100 inch).

291. To remove the lift pump from most tractors, except those equipped with only single speed (540 rpm) ground speed pto, it is necessary to drain transmission oil and remove lift cover as outlined in paragraph 285. Split tractor between differential housing and transmission housing as outlined in the appropriate paragraph in the transmission section. Remove pto output shaft and cover from right side of rear axle center housing. Disconnect hydraulic lines and remove pump supports (S—Fig. 293), then remove hydraulic pump unit through front of differential housing.

Installation is reverse of removal procedure. Always install new split pins, "O" rings and gaskets. Carefully locate the pump on ipto clutch and install supports (S). Tighten nuts or screws retaining support (S) to right side, but leave left side support retaining nuts or screws loose until tractor assembly is complete. Be sure that engine will not start and crank engine with starter, then tighten nuts or screws

Fig. 293—Hydraulic pump supports (S) are attached to each side of rear axle center housing.

retaining support to left side of axle center housing. This procedure will ensure correct alignment of shafts.

292. OVERHAUL. To disassemble the removed pump, first clean the outside of pump, then unbolt and remove the pressure relief valve assembly (1—Fig. 294). Remove clip (2) and lever (3). Remove the four lower screws (4) retaining cover (7), then separate inlet housing (17) from front section of pump. Remove the remaining two screws (6) and separate rear cover (7) from inlet housing. Remove clip (9), nut (10), spring (11), washer (12), "O" ring (13), filter (14), inner shroud (15) and upper "O" ring (16). Unhook spring (18), remove screws (19), then remove response control lever (20) and retainer (21). Ball, spring and bushing (24) can be removed after removing response lever. Compress spring and remove retainer snap ring (25) retainers (26 and 28), spring (27), seal (29) and valve (30).

Remove cap (31—Fig. 295), compress spring (37) using MF352 retainer clip or equivalent and remove retaining ring (32). Withdraw control valve (33) and balancer tube (34). Be careful that spring (37) is not ejected and remove special tool MF352. Remove collar (35), guide (36), spring (37) and square disc (38). Remove four retaining nuts (39), then withdraw front cover (40) from studs. Remove front thrust washer (42) and "O" rings (41). Slide valve chambers (43) from studs, withdrawing pistons (44 and 50), cam blocks (45 and 49) and oscillator drive (46). Carefully separate parts and remove camshaft (51) from rear body (52). Control valve body (53) can be driven from rear body if new "O" rings and backup rings are to be installed. Special tool (MF351 or equivalent) can be used to remove valve assemblies from valve chambers (43) if service is required.

Fig. 294—Rear view of partially disassembled hydraulic lift pump filter, rear cover and inlet housing. Refer to Fig. 295 for pump parts at front.

1. Pressure relief valve & housing
2. Clip
3. Actuating lever
4. Screws (4 used)
5. Gasket
6. Screws (2 used)
7. Rear cover
8. Gasket
9. Clip
10. Nut
11. Spring
12. Washer
13. "O" ring
14. Filter
15. Inner shroud
16. "O" ring
17. Inlet housing
18. Spring
19. Screws (2 used)
20. Response control lever
21. Retainer
22. Bushing
23. Washer
24. Bushing, spring & ball
25. Snap ring
26. Retainer
27. Spring
28. Retainer
29. Seal
30. Valve

When assembling, backup rings are outside "O" rings at positions (54—Fig. 295). Special tool (MF353 or equivalent) can be used to install backup rings and "O" rings (54) on control valve. Align pin for control valve (53) when installing and use special tool (MF354 or equivalent) to drive valve into rear body bore.

Valve seats can be recut in chambers (43) using special tool (MF349 or equivalent). Special tools (MF350, MF351 and MF353) can be used to assemble valves and valve chambers.

Reassemble by reversing disassembly procedure. Lubricate all parts liberally before assembling and always install new "O" rings and retaining rings. Chamfered corners (C) of piston yokes face away from each other. Tighten nuts (39) to 40 N•m (30 ft.-lbs.) torque. If only two of the nuts (39) are special type, install these two on right top and bottom left studs. Assemble balancer tube (34) to control valve (33), lubricate valve and insert into rear body.

Tighten nut retaining relief valve manifold (1—Fig. 294) to 27 N•m (20 ft.-lbs.) torque. Tighten screws (4 and 6) evenly to 20 N•m (15 ft.-lbs.) torque, while checking to be sure that control valve (33—Fig. 295) continues to move freely. After clip is fitted to pin (47), bend end over to prevent its removal.

AUXILIARY HYDRAULIC PUMP

All Models So Equipped

293. The power steering pump of most models is tandem or dual type, incorporating one section for power steering and the other section for the auxiliary hydraulic system. Refer to paragraph 25 for removal and to paragraph 30 for overhaul.

AUXILIARY SYSTEM PRESSURE

All Models So Equipped

294. A pressure maintaining valve (5—Fig. 296) is fitted to tractors with ipto and is used to maintain sufficient pressure to operate this clutch. Valves with different maintaining pressures are used for tractors with three-cylinder engines 1700-2000 kPa (250-290 psi) than for tractors with four or six cylinder engines 1900-2200 kPa (275-325 psi). Pressure is stamped on valve body and should be correct for installation. Pressure is not adjustable, but can be checked at port (8) with oil warm and engine operating at 1200 rpm. System pressure may be increased by clogged hydraulic system filter, so new filter should be installed

Fig. 295—Hydraulic lift pump consists of four pistons (44 and 50) contained in cylinders of valve chambers (43). Chamfered corners (C) of piston yokes face away from each other.

31. Cap	38. Square disc	45. Cam block	52. Rear body
32. Retainer ring	39. Nuts (4 used)	46. Oscillator drive	53. Control valve body
33. Control valve	40. Front cover	47. Pin	54. "O" rings
34. Balancer tube	41. "O" rings (4 used)	48. Oscillator	& backup rings
35. Collar	42. Thrust washers	49. Cam block	55. Spacer
36. Guide	43. Valve chambers	50. Piston	56. Washer
37. Spring	44. Piston	51. Camshaft	

Fig. 296—Views of installed pressure maintaining valve (5). Cross section shows internal components of pressure maintaining valve. Three and six cylinder models are not equipped with Multi-Power and will not be equipped with solenoid (2), block or outlet at (3).

1. Pressure from steering system	5. Pressure maintaining valve	7. Low pressure switch	10. Dampening shuttle
4. Manifold	6. Pressure to auxiliary hydraulic pump	8. Test port	11. Poppet
		9. Spring	12. Orifice

Fig. 297—Exploded view of relief valve (2) for auxiliary hydraulic system. Pressure can be checked at port for plug (1). Shims (3) are located between spring and plug.

before checking pressure. Low pressure could be caused by:

1. Faulty pressure maintaining valve.

2. Steering pump worn

3. Leakage of ipto system caused by loose connection or ipto brake cylinder leaking.

295. Auxiliary system pressure relief valve (Fig. 297) should maintain auxiliary system pressure at 17,000-19,000 kPa (2500-2750 psi) for all models. Pressure can be checked by attaching appropriate test gauge in place of port plug (1) located in manifold mounted just in front of left rear axle housing. Start engine and make sure fluid is warmed. Be sure that selector valve is in linkage position and set engine speed to 1200 rpm. Pull one of the auxiliary control valve spools back to lift position and note pressure indicated by the installed test gauge. Low pressure could be caused by:

1. Auxiliary pump worn.

2. Relief valve faulty.

3. Internal leakage across spool valve.

Low relief valve setting can be corrected by removing plug and adding shims (3) as required. Be sure to install new seals and tighten relief valve plug to 70-100 N·m (52-74 ft.-lbs.) torque.

296. Auxiliary system flow can be checked by attaching an appropriate flow tester to outlet port connection. Direct fluid from tester back to the transmission reservoir through filler opening. Start engine, make sure that flow meter restrictor valve is open and allow fluid to reach normal temperature. Make sure that ipto is in "OFF" position and set engine speed to 2000 rpm. Slowly close flow meter

restrictor valve until pressure increases to 13,800 kPa (2000 psi) and observe volume of flow. Open restriction as soon as flow is determined. For tractors with three-cylinder engine, new pump should flow 31 L/minute (8.2 gal./minute) or at least 27 L/minute (7.1 gal./minute). New pumps on six cylinder tractors should deliver 32 L/minute (8.4 gal./minute) or at least 28 L/minute (7.4 gal./minute) for used pumps. Low volume of flow could be caused by worn pump, blocked (or partially blocked) suction strainer or insufficient amount of oil in transmission sump.

AUXILIARY REMOTE CONTROL VALVES

All Models So Equipped

297. One, two or three auxiliary spool valves may be installed on bracket at rear of tractor. Cap parts (A—Fig. 298) are for valves with detent positions which return (kickout) to neutral when cylinder reaches end of stroke. Cap parts (B) are for valves with spring return to neutral. Cap parts (C) provide

Fig. 298—Exploded and cross section of auxiliary spool control valve. Cap parts (A) are for valves with three position detent, cap parts (B) are for valves with spring return to neutral and cap parts (C) are for valves with detent type with float position.

four positions, including detent and float positions. Additional special valves are also available including type without kickout and valve for use with hydraulic motors. Service is conventional and should be accomplished if seals and repair parts are available. "Loctite 542" or equivalent should be applied to threads of shoulder bolt (1 or 2) and tighten shoulder bolt to 7 N·m (5 ft.-lbs.) torque. Apply "Loctite 222" to threads of screws (3, 4 or 5) and tighten to 7 N·m (5 ft.-lbs.) torque. Tighten change-over valve (6) to 22 N·m (17 ft.-lbs.) torque.

To adjust kickout relief pressure of valves (C), attach inlet hose for flow meter to valve and connect outlet side of flow meter to the other port of valve. Remove plug (7), open flowmeter fully and start engine. After fluid reaches normal operating temperature, set engine speed at 2000 rpm and move valve control to locked position. Close the flow meter restriction valve slowly until pressure causes valve to kickout (return to neutral). Kickout pressure should be 17,200 kPa (2500 psi) and can be changed by turning adjusting screw (8) with a 5/32 inch Allen wrench through opening for plug (7). Turning screw clockwise will increase kickout pressure and each 10° will change pressure about 200 kPa (30 psi). Check setting several times before reinstalling plug (7).

NOTES

NOTES

NOTES

NOTES